Surik Khudaverdyan, Ashok Vaseashta
Semiconductor Photodetectors

Also of interest

Fiber-Based Optical Resonators.
Cavity QED, Resonator Design and Technological Applications
Deepak Pandey, 2024
ISBN 978-3-11-063623-9, e-ISBN (PDF) 978-3-11-063626-0,
e-ISBN (EPUB) 978-3-11-063630-7

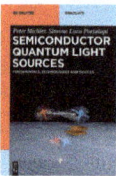

Semiconductor Quantum Light Sources.
Fundamentals, Technologies and Devices
Peter Michler, Simone Luca Portalupi, 2024
ISBN 978-3-11-070340-5, e-ISBN (PDF) 978-3-11-070341-2,
e-ISBN (EPUB) 978-3-11-070349-8

Photonic Reservoir Computing.
Optical Recurrent Neural Networks
Daniel Brunner, Miguel C. Soriano, Guy Van der Sande (Eds.), 2019
ISBN 978-3-11-058200-0, e-ISBN (PDF) 978-3-11-058349-6,
e-ISBN (EPUB) 978-3-11-058211-6

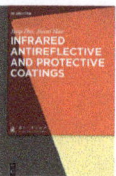

Infrared Antireflective and Protective Coatings
Jiaqi Zhu, Jiecai Han, 2018
Together with: National Defense Industry Press
ISBN 978-3-11-048809-8, e-ISBN (PDF) 978-3-11-048951-4,
e-ISBN (EPUB) 978-3-11-048819-7

Surik Khudaverdyan and Ashok Vaseashta

Semiconductor Photodetectors

Optical Spectrometry

DE GRUYTER

Authors
Prof. Dr. Surik Khudaverdyan
Armenian National Polytechnic University
Teryan St 105
Yerevan
Armenia
xudaver13@mail.ru

Prof. Dr. Ashok Vaseashta
International Clean Water Institute
9108 Church Street, P.O. Box 258
Manassas, VA 20108
USA
prof.vaseashta@ieee.org

ISBN 978-3-11-142787-4
e-ISBN (PDF) 978-3-11-142802-4
e-ISBN (EPUB) 978-3-11-142831-4

Library of Congress Control Number: 2024936432

Bibliographic information published by the Deutsche Nationalbibliothek
The Deutsche Nationalbibliothek lists this publication in the Deutsche Nationalbibliografie;
detailed bibliographic data are available on the internet at http://dnb.dnb.de.

© 2024 Walter de Gruyter GmbH, Berlin/Boston
Cover image: aislan13/E+/Getty Images
Typesetting: Integra Software Services Pvt. Ltd.

www.degruyter.com

Foreword

Photodetectors are devices that detect light and convert it into electrical signals. Such devices are ubiquitous in our daily lives, and some of the applications include digital cameras, smartphones, solar cells, fiber optical communication, barcode scanners, smoke detectors, automotive, biomedical imaging, and biometric security systems. These are just a few of many examples that highlight the pervasive use of photodetectors in our daily lives, indicating their tremendous importance. Using cyber-physical systems (CPS), our surrounding environment is now increasingly connected through smart and intelligent systems. In conjunction with commercial photosensors, the advances in CPS have resulted in several experimental platforms that provide Internet of things (IoT) capabilities. With an exponential increase in IoT devices, the scaling potential of monolithic integration of photonic-electronic devices is likely to surpass the transistor-era progress of the twenty-first century.

Furthermore, by providing a pathway for enormous bandwidth and interconnection hierarchy, distributed systems can reduce functional latency, in addition to cost and energy requirements. In conjunction with green IoT and artificial intelligence-based IoT, there is a parallel effort to create highly sensitive devices by device design to conserve energy. So far, silicon-based platforms have provided the materials, tools, and design experience to define gaps and solutions for common, cross-market applications. However, numerous applications require deciphering information from very weak optical signals, such as from radiation, medical imaging, industrial nondestructive testing, quantum technologies, astronomy, and various routine measurements. Hence, it is necessary to design photodetectors with high photosensitivity using various technological innovations to reduce the noise level, such as with two inversely directed barriers, as described in this book by the authors, in which the currents of devices mutually compensate each other and create low dark current, resulting in high photosensitivity thresholds. The implementation of internal amplification of photocurrents in such devices provides high photosensitivity.

This book presents the mechanism for the injection amplification of the photocurrent in devices based on cadmium telluride and silicon with a high-resistance sublayer, as well as the study of creating highly sensitive devices that are resistant to radiation in optical and X-ray ranges of electromagnetic waves. Particular attention is drawn to the mutual compensation process for photocurrents arising in opposite potential barriers covering the layer during longitudinal absorption of radiation in the sublayer. Using structures on the base cadmium telluride and silicon, as an example, the phenomenon of a change in the sign of the spectral photocurrent and the capability of wave measurement is provided using this phenomenon.

The global spectral analysis market is currently focused on developing semiconductor photodetectors with spectral-selective sensitivity for spectral analysis. Using such a photodetector in spectrometry will eliminate the use of optical-mechanical systems due to the new physical principle described in this book, which will ensure high

https://doi.org/10.1515/9783111428024-202

resolution and reliability of spectrum recording. This is particularly useful for remote spectral analysis, identification, and assessment of substances in air, water, and food, assessment of the effects of substances on humans, animals, and vegetation, and detection and elimination of pollution sources. Needless to mention here, the spectral analysis of the electromagnetic radiation transmitting the information from the object with the help of highly sensitive sensors is extremely essential.

It is imperative that the ever-expanding workforce needs books, reference manuals, and a comprehensive source of information for this specialized field. Typically, such information is available in several books, reference guides, and literature. It is quintessential to have the knowledge base in one consolidated book, such as this. There are competing texts but not to the extent as presented and covered by this book. Some books describe similar contents but not in their entirety, as is presented in this book. Likewise, many books on semiconductor physics describe the basic mechanism in chapters but not with as much detail and precision as is provided in the book. Hence, this is likely to serve as text, research manual, and source of information to the researchers working in this field.

The authors, Prof. Khudaverdyan and Prof. Vaseashta, have done an excellent work in this book, collating the work done over the past 25 years. The concepts, theoretical basis of novel ideas, and devices, as reported in this book, will likely guide many scientists and student researchers in the coming years. On a personal note, we have known Prof. Khudaverdyan, as a colleague for many years, at the National Polytechnic University of Armenia. Prof. Vaseashta and Prof. Khudaverdyan organized a highly successful NATO Advanced Research Workshop in Yerevan in 2012. We had the opportunity to read several of his articles, which proved to have an immeasurable benefit for our graduate students, engineers, and scientists, and provided comprehensive guidance toward the current state-of-the-art science and technology.

We wish the authors a big success with this book and are sure that many students and researchers will be reading this book with great interest.

Prof. Dr. Areg Grigoryan
Vice-rector for Science
Armenian National Polytechnic University, Armenia

Prof. Dr. Armine Avetisyan
Head, Center for Scientific Research and Innovation Development
Armenian National Polytechnic University, Armenia

Preface

The discovery of the photoelectric effect revolutionized our understanding of the nature of light and set the foundation for quantum mechanics. In 1887, Heinrich Hertz first observed that light could eject electrons from a metal surface. In 1905, Albert Einstein provided a theoretical explanation for this phenomenon. Photodetectors exploit the photoelectric effect to detect and measure light in many applications beneficial to society. However, numerous applications require deciphering information from very weak optical signals, such as from radiation, medical imaging, industrial nondestructive testing, quantum technologies, astronomy, and various other routine measurements. Hence, it is necessary to design photodetectors with high photosensitivity using various technological innovations to increase signal/noise ratio. With this as the background, this monograph is written, in particular, based on research conducted on a fundamental mechanism for the injection amplification of the photocurrent in photodetectors. The samples were based on cadmium telluride (CdTe) and silicon with a high-resistance sublayer for creating highly sensitive devices, especially selected regions of the electromagnetic spectrum. Particular attention is drawn to the mutual compensation process for photocurrents generated in the oppositely directed potential barriers covering the base during the longitudinal absorption of the radiation in the sublayer. The photoelectronic processes occurring in these semiconductor structures are investigated, and the expressions connecting the parameters of the optical radiation and the structure are obtained. The algorithm developed using these expressions is based on a new spectral analysis mechanism which is realized in inexpensive, small-size, with reduced material requirements, and energy-saving devices. All this is considered in terms of solving urgent problems of the quantitative remote identification of the components of an optically transparent medium.

One of the relevant directions of this research is also to solve the problem of the development of a semiconductor structure in which the electronic processes provide the high-accuracy spectral analysis of electromagnetic radiation. Most of the photospectrometers currently available in the market have several disadvantages that limit their use in certain areas and hinder their widespread use, such as they require periodic adjustments, lack of reliability, size, weight, cost, lack of flexibility, and need for an external unit to control, process, and display the spectral data. Hence, it is crucial to create highly sensitive sensors for several optical signals and to conduct spectral analysis to extract relevant information signals.

The objective of this monograph is to investigate highly sensitive photodetector structures by the process of injection amplification. This book starts with a brief overview of the importance of highly sensitive photodetectors and how using device physics and design engineering can achieve certain configurations to achieve the desired sensitivity, more specifically, by introducing a high-resistivity layer in CdTe-based devices, where, in fact, most of the photocurrent occurs. The book goes on to discuss the mechanisms, operation, and theory of the process of injection amplification in struc-

https://doi.org/10.1515/9783111428024-203

tures with a high-resistivity layer. The devices can be customized to different wavelengths, such that the structural features can even allow the Earth's background radiation to be detected. Next, the book provides a theoretical basis for the functionality of structures with a high-resistivity layer based on CdTe and silicon. The mechanism for changing the sign of the spectral photocurrent and its significance is provided. The next chapter describes several structures with selective spectral sensitivity and their common drawbacks. A new structure is proposed as an alternative, in which spectral selectivity in the new structure is studied using the new mechanism. This leads to the creation of small-sized, efficient, and reliable spectral analysis instruments. The next chapter describes the relevance of developing small-sized, inexpensive sensors with high spectral sensitivity, suitable for remote identification in field conditions, and the development of algorithms for accurately recording information using such devices. The dependence of spectrophotometric properties on the structural and technological parameters of the structure is described. The next chapter describes the process of mutual compensation of photocurrents of opposing potential barriers in device structures, thus providing low dark-current values and photosensitivity thresholds. The mathematical modeling of photoelectronic processes is explained, which underlies the algorithmic operation for obtaining the spectral dependence of the intensity under optical radiation. The next chapter provides an in-depth discussion of the results of photospectrometric studies of photodetectors. The spectral dependencies obtained using single- and multicolor radiation sources, identical to the reference sources, indicate the promise of the structures under investigation. The next chapter describes various mechanisms, such as the effect of tunneling on the spectral response and the selective spectral photosensitivity of the samples under study. The nature of the relationship between the energy parameters of absorbed waves and structural parameters has been identified. Additionally, using mathematical modeling, the algorithm developed for obtaining the spectral dependence of intensity resulted in the implementation of a program for the transition from absorption coefficient to wavelength. Structures with injection of carriers through a forward-biased n-p junction and without injections have been studied. The conditions for changing the sign of the spectral photocurrent and long- and short-wave maxima were described. The next chapter determines the voltage range at which the sign of the spectral photocurrent is inverted when depleted regions merge. It is shown that this range is comparable to the difference in heights of potential barriers. The near-surface barrier has the advantage of absorbing quanta and creates a short-wavelength spectral maximum of the photocurrent while counteracting the rear barrier in the long-wavelength part of the spectrum. The difference in the height of the potential barriers leads to a change in the sign of the spectral photocurrent both in the absence of a bias voltage and for voltages that increase the height of the low (reverse-biased) barrier to its comparison level. Outside of this voltage range, the sign does not change. The last chapter presents that under conditions of longitudinal illumination, spectral distributions of photosensitivity with short-wave and long-wave maxima were identified. It is observed that the numerical values of photosensitivity are unusually high. Furthermore, there is an internal enhancement of the photocurrent, which is explained by the injection of photogenerated carriers through the forward-biased potential barrier, which leads to a decrease in the height of both potential barriers.

This monograph will be useful for professionals and researchers working in the fields of materials science and engineering, physics, device modeling, sensors, detectors, and various other related disciplines, namely, environment, chemistry, toxicology, and subjects related to public policies and decision support. Additionally, the book will be of interest to scientists/researchers, graduate students, and specialists. It may also be used in conjunction with a textbook and as an assignment for independent study courses. In addition, it is intended for commercial markets that specialize in products, such as highly sensitive and wavelength-selective semiconductor photodetectors, optical spectral analyses, highly sensitive photodetectors with low thresholds, and devices for researchers involved in photonics. The exponential increase of the Internet of things will continue to revolutionize the need for highly sensitive devices for efficient sensing and simultaneously conserve power.

The authors acknowledge the contributions of many fellow scientists, students, and reviewers who assisted in providing their input, remarks, and critical assessment of the subject matter. Such feedback was critical to the quality of this book. The authors immensely appreciate the funding, in part, by the Science Committee of the Republic of Armenia for research (project no. 21T-2B028). The authors immensely appreciate the publisher, De Gruyter, Ms. Melanie Götz, and Ms. Helene Chavaroche for their guidance, patience, and help with this book.

The authors, in no way, claim that the contents of this book completely address the challenge of the scientific issue presented in this book. In fact, in subsequent editions, the authors will address additional ongoing and upcoming challenges and opportunities.

Disclaimer: The opinions, recommendations, and propositions presented in various chapters represent the findings, views, and opinions of the authors and may not represent the position(s), views(s), and/or opinion(s) of their institutions. The authors do not promote, endorse, or express their dissent against any product(s) that may have been mentioned in the chapters. Although due precautions are exercised before publication, the authors, however, do not take any responsibility for unintentional infringement of any material of either their own or from elsewhere. For all suggestions, corrections, or exchanges of information, the reader is advised to contact the authors directly.

Prof. Dr. Surik Khudaverdyan
Yerevan, Armenia

Prof. Dr. Ashok Vaseashta
Manassas, VA, USA, and Chisinau, Moldova

Contents

Part 1: **Photodetector structures**

Part 4: **Measurement methods and photoelectronic characteristics**

Part 5: **Research results**

Part 6: **On the semiconductor spectroscopy**

Glossary of symbols and scientific notations

Acronyms

APD	Avalanche photodiodes
CCD	Charge-coupled devices
CdTe	Cadmium telluride
CMOS	Complementary metal oxide semiconductors
CVC	Current-voltage characteristic
DC	Direct current
DPB	Double potential barrier
FET	Field-effect transistor
FIR	Far-infrared
FTIR	Fourier-transform infrared spectroscopy
GaAs	Gallium arsenide
HgCdTe	Mercury cadmium telluride
IC	Integrated circuit
InGaAs	Indium gallium arsenide
InGaN	Indium gallium nitride
IR	Infrared
LAC	Lux-ampere characteristic
LED	Light-emitting diode
MBE	Molecular beam epitaxy
MOS	Metal oxide semiconductor
NDT	Nondestructive testing
NEP	Noise equivalent power
PMT	Photomultiplier tube
RF	Radiofrequency
SEM	Scanning electron microscopy
Si	Silicon
SiC	Silicon carbide
SNSPD	Superconducting nanowire single-photon detectors
SP	Semiconductor photodetector
SPADS	Single-photon avalanche diodes
T	Temperature (typically in kelvin)
TiW	Titanium tungsten
UV	Ultraviolet
XRD	X-ray diffraction
ZnO	Zinc oxide

Scientific notations

E_n	Absorbed energy
α	Absorption coefficient
A	Ampere
a.u.	Arbitrary units

https://doi.org/10.1515/9783111428024-205

E_g	Bandgap
φ_0	Barrier of the p-n junction
k	Boltzmann's coefficient
cm	Centimeter
I	Current
I_D	Dark currents
N_c	Effective density of state of the conduction band
n	Electron concentration
$\mu_n\,\mu_p$	Electron and hole mobility
q	Electron charge
eV	Electron voltage
ΔF_n	Fermi level for electrons
ΔF_p	Fermi level for holes
Hz	Frequency
ρ_I	High-resistance layer
K	Kelvin
L	Length of the diffusion
τ	Lifetime of minority carriers
I_L	Light current
μm	Micrometer
μV	Microvoltage
μW	Microwatt
n	Nano
nA	Nanoamperes
N_{Fot}	Number of absorbed photons
n_n	Number of main electron carriers
N_e	Number of photoelectrons
I_{Ph}	Photocurrent
h	Planck's constant
Γ	Quantum efficiency
r.u	Relative units
ρ_b	Resistivity of the base region
s	Second
D_{Ag}	Silver diffusion coefficient.
$S_{i\gamma}$	Spectral dose sensitivity
c	Speed of light
D_{Ag}	The silver diffusion coefficient
D	Thickness of the base region
V	Voltage
W	Watt
ϑ	Wave frequency
λ	Wavelength

Introduction

The history of photodetectors dates back to the nineteenth century when experiments with light-sensitive materials led to the discovery of the photoelectric effect by Heinrich Hertz in 1887. However, it was not until 1905 when Albert Einstein provided a theoretical explanation of the phenomenon that the foundation for modern photodetection was laid. In the early twentieth century, selenium photodetectors emerged as one of the first practical devices capable of converting light into electrical signals. Over the years, advancements in semiconductor technology fueled the development of more efficient and versatile photodetectors, including photodiodes, phototransistors, and avalanche photodiodes (APDs), enabling applications ranging from telecommunications to imaging and sensing. Today, photodetectors play a vital role in various fields such as astronomy, remote sensing, medical imaging, and consumer electronics.

Highly sensitive photodetectors are devices designed to detect even extremely low levels of light with high precision and efficiency. They play a critical role in various applications, where sensitivity to light is paramount, such as in astronomy, environmental monitoring, medical imaging, and telecommunications. These photodetectors typically utilize advanced semiconductor materials, such as silicon (Si), indium gallium arsenide (InGaAs), or mercury cadmium telluride (HgCdTe), which exhibit high quantum efficiency and low-noise characteristics. Developing highly sensitive photodetectors involves a multidisciplinary approach that encompasses materials science, device engineering, and fabrication techniques. Researchers focus on utilizing advanced semiconductor materials with high quantum efficiency and low-noise characteristics, such as Si, InGaAs, or HgCdTe. They also explore novel materials like perovskites and quantum dots for their potential to enhance sensitivity. Tailoring the design and structure of photodetectors, including optimizing the active region size, doping profiles, and surface passivation, is crucial for maximizing light absorption and minimizing noise sources. Moreover, incorporating innovative concepts such as APDs, single-photon avalanche diodes, or superconducting nanowire single-photon detectors further enhances sensitivity. Continuous refinement of fabrication processes, including precise lithography, doping, and deposition techniques, is essential for achieving reproducible and reliable devices. Collaborations between researchers from various disciplines drive progress in developing highly sensitive photodetectors, facilitating advancements in fields such as astronomy, medical imaging, and quantum information processing.

The focus of this book is on the fundamental mechanism for the injection amplification of the photocurrent in samples based on cadmium telluride (CdTe) and Si with a high-resistance sublayer, as well as on the research on the possibilities of creating highly sensitive devices in the optical and X-ray ranges of electromagnetic waves. Particular attention is drawn to the mutual compensation process for photocurrents generated in the oppositely directed potential barriers covering the base during the longitudinal absorption of the radiation in the sublayer. Using structures based on Si

https://doi.org/10.1515/9783111428024-001

and CdTe, the phenomenon of a change in the sign of the spectral photocurrent and the possibilities of wave measurement provided by this phenomenon are demonstrated. The photoelectronic processes occurring in these semiconductor structures are investigated, and the expressions connecting the parameters of the optical radiation and the structure are obtained. The algorithm developed using these expressions is based on a new spectral analysis mechanism which is realized in the form of inexpensive, small-size, with reduced material requirements, and energy-saving prototype devices. Furthermore, the application of the elaborated mechanism is suitable for solving urgent problems of the quantitative remote identification of the components of an optically transparent medium.

The importance of applied science has greatly increased in the current stage of the society's development. Mankind has to respond to new challenges: the problems connected with human health, biological and environmental security, water monitoring, food control, climate changes, space exploration, the study of the Earth's interior, the raising of a powerful army, and so on. Most of these problems are solved with the help of sensors that record and analyze the initial information. Thus, it is crucial today to create highly sensitive sensors for different optical signals and to carry out the spectral analysis of optical information signals. The primary sensor is a semiconductor photodetector (SP). To make highly sensitive sensors with new functionality, it is necessary to increase the reliability, spectral-selective sensitivity, and speed of the SP and to reduce its cost, weight, size, energy, and material consumption as compared to the existing ones.

The spectrum analysis market worldwide focuses on the development of an SP with spectral-selective sensitivity to carry out the spectral analysis. The use of such a photodetector in spectrometry will eliminate the use of optical-mechanical systems due to the new physical principle used in it and will ensure high resolution and reliability of the spectrum registration. The photodetector will be small, low cost, and high speed. Furthermore, it will yield useful and highly desired applications such as the remote spectral analysis, the remote identification, the remote quantity assessment of hazardous substances in air, water, and food, the assessment of the impact of hazardous substances on humans, animals, and vegetation, and the detection of pollution sources.

Since the threat to the environment is becoming more and more unpredictable, there is a growing need for the development of user-friendly remote monitoring systems that will enable the remote analysis of situations in different environments. Such systems should have certain properties, high sensitivity, and selectivity, as well as many inputs for analyzing various dangers. There is also a need for the development of systems that will enable remote monitoring, analyzing, and on-screen displaying of the data in the environment utilizing modern communication systems. Hence, the spectral analysis of electromagnetic radiation transmitting the relevant information from sources with the help of primary sensors is essential.

The modern spectrophotometric systems lack universality. To perform a new function, additional equipment, computer, and software support are required. It makes these systems expensive and difficult to use in field conditions. In addition, the analysis is accompanied by the quantitative and qualitative scatter of the obtained data. The reliability of the results is directly related to the parameters of the equipment in use and requires the improvement of measuring devices and the development of more rational methods of the analysis of the experiment results. Besides, the more information is received, the more difficult it is to interpret the measurement results. Thus, the development of a more complete algorithm of useful information is required.

Most of the photospectrometers currently available on the market have several disadvantages, which limit their use in certain areas and hinder their widespread introduction into new areas of life (e.g., food and water safety testing). These disadvantages include:
– the application of complex and precise mechanical and optical units that require periodic adjustment;
– lack of high reliability;
– size and weight;
– cost;
– the necessity for an external unit to control, process, and store the measurement results and display the spectral data on the screen; and
– low flexibility trade-off, which complicates the enhancement and the possibilities of new applications.

Hence, one of the most relevant directions to solve this challenge is the development of semiconductor structures in which the electronic processes provide the high-accuracy spectral analysis of electromagnetic radiation. Under the current circumstances, the above-stated disadvantages limit the application of spectrophotometers. They cannot be used for the remote analysis carried out over large areas in field conditions, which is in high demand today. Thus, there is a need for low-cost, high-speed, small-size, portable, high-sensitivity primary sensors equipped with signal processing, signal localization, and telecommunication units. To address the current challenges, it is necessary to discover and apply the functional properties of specially designed two-barrier SP structures of the photospectrometric devices. The structures ensure high-accuracy remote registration and have spectral-selective sensitivity. The solutions to the above-stated problems are reflected in this book.

Part 1: **Photodetector structures**

Chapter 1
Photodiode structures with a high-resistance layer

1.1 Introduction: injection amplification of photocurrent in CdTe-based diode structures

The advances in the field of improvement of solid-state photoreceivers have attained a high level of stage due to advances in materials synthesis, and device design and processing. In the foreground, there are problems related to search for new materials and physical-technical principles capable of providing high photosensitivity and radiation resistance and enhancing the functionality of photodetectors. Due to these advances, photodetectors are widely used both as discrete photoelectronic devices and as active elements in two- and three-dimensional photoelectronic integrated circuits.

The problems with high photosensitivity and radiation resistance can be solved by using a structure based on A^2B^2-type high energy gap semiconductors. The large width of the energy bandgap provides for low thermal generation of charge carriers and a low noise level at room temperature. The lifetime of nonequilibrium charge carriers in these compounds can reach the values of 10^{-4}–10^{-2} s, which makes them highly sensitive to radiation. A large average atomic number and a high concentration of intrinsic defects provide for the effective absorption of penetrating radiation and high radiation resistance.

The creation of structures with internal photocurrent amplification, based on high energy gap semiconductors, makes it possible to obtain an ultra-linear current-voltage characteristic (CVC) and to ensure high photosensitivity at a low noise level. In excitation, an additional carrier injection from contacts is realized in structures. The additional injection can be realized with a high-resistance layer placed between two potential barriers. Cadmium telluride (CdTe) is the only semiconductor from A^2B^2-type high energy gap semiconductors that is slightly inclined to self-compensation. It is easy to obtain CdTe-based semiconductors with both n- and p-type conductivity and to create an n-p junction in them. This chapter describes a theoretical basis for the experimental investigation of photodetector structures with a high-resistance CdTe-based layer, in which an injection amplification of the photocurrent occurs.

1.1.1 Principle of photocurrent amplification

CdTe is the most suitable material to produce a highly sensitive and radiation-resistant photodetector and a high particle detector (Cola et al., 2012; S. K. Khudaverdyan, 2003; Mohammadi et al., 2009). The idea of the injection amplification in a "long diode" with a semiconductor thickness that is several times larger than the length of the diffusion bias of minority carriers was first proposed by V. I. Stafeev and realized in classical

https://doi.org/10.1515/9783111428024-002

semiconductors, namely, silicon and germanium (S. K. Khudaverdyan et al., 2014; I. M. Vikulin et al., 2008).

In case of an abrupt nonsymmetrical n-p junction, when the generation and the recombination in the space charge region can be neglected and the current through the junction can be considered purely hole, the solution of the system of the diffusion-drift equations leads to the following expression for the CVC of the diode in the high injection-level mode (I. M. Vikulin et al., 2008; I. Vikulin & Stafeev, 1990):

$$I = I_c \exp\left(\frac{qV}{ckT}\right) c = 2 \times \left[b + ch\left(\frac{d}{L_p}\right)\right] / (b+1) \tag{1.1}$$

where d is the thickness of the base region (semiconductor thickness), L_p is the length of the diffusion bias of minority carriers, and V is the total voltage drop on the diode:

$$I_p = \left[D_p \tau_p \left(\frac{2b}{b+1}\right)\right]^{1/2} \tag{1.2}$$

When deriving eq. (1.1), the rear contact is considered to be ohmic. Equation (1.1) shows that if $d/L_p > 1$, there is a very strong dependence of the current on the change of the lifetime or the mobility of nonequilibrium carriers. These parameters can change with the change of the current flowing through the diode or under external influences (radiation, magnetic field, etc.).

The operating principle of photodetectors with $d/L_p > 1$ is described further. The carriers injected into the base of the photodetector modulate its conductivity. With the constant voltage applied to the diode, an increase in the base conductivity leads to the redistribution of voltage between the base and the n-p junction. The voltage drop on the n-p junction increases, which leads to an increase in the injection of minority carriers through the junction. An increase in the injection, in turn, further modulates the base conductivity. There occurs a new redistribution of voltage. If this increases the lifetime or mobility of carriers in the base region, even the change of their values several times can lead to an increase in the current through the diode by several orders of magnitude (I. M. Vikulin et al., 2008; I. Vikulin & Stafeev, 1990). The virgin single crystal had indium-doped n-type conductivity, a resistivity of $10^7 - 10^8 \mathrm{Ohm} \times \mathrm{cm}$, and a thickness of $400\mu m$. The p-region was created by thermal diffusion of silver. Figure 1.1 presents the design of the photodetector structure (S. Khudaverdyan et al., 2005; S. K. Khudaverdyan, 2003).

1.1.2 Current-voltage characteristics of diodes

At room temperature, CVCs of these diodes (Figure 1.2) in the dark were slightly different from linear ones (curves 1 and 2).

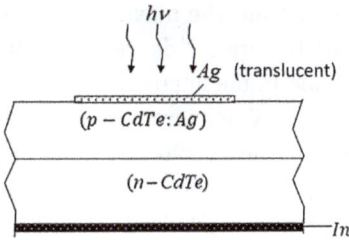

Figure 1.1: The design of the photodetector structure based on CdTe.

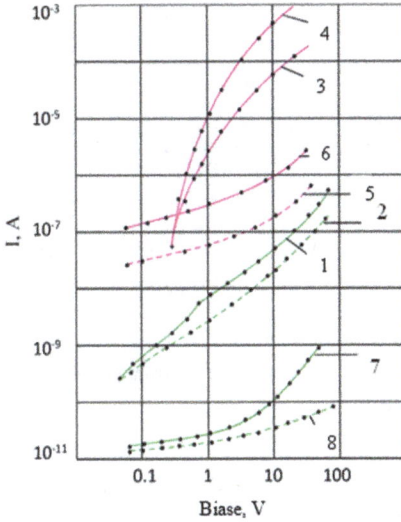

Figure 1.2: CVC of diode structures in the dark (curves 1, 2, 7, and 8) ånd under illumination (curves 3–6). Incident radiation power (V), μW: 0.1 (curves 3 and 5), 0.5 (curves 4 and 6); $T = 300$ K (curves 1–6); $T = 80$ K (curves 7 and 8); curves 1, 5, 4, and 7 – forward bias; curves 2, 5, 6, and 8 – reverse bias.

The rectification factor was small and was approximately equal to 3. At liquid nitro-gen temperature, the dark currents decreased abruptly. At the same time, the direct and inverse branches became sublinear at voltages up to 10 V. While the inverse branch of the CVC remained sublinear at high voltages, the direct branch had a no-ticeable ultra-linearity (curves 7 and 8), at which point the rectification factor also increased.

1.1.3 Spectral characteristics of diodes

The spectral dependencies of diodes were studied when they were illuminated with monochromatic light. The beam was obtained from an incandescent lamp, by passing through a monochromator YM-2. Figure 1.3 presents the measurement results of sam-ples at different bias voltages and incident radiation intensities.

In this figure, along the ordinate axis, the relative values of spectral photosensi-tivity are given. The light intensity is changed by the voltage of the lamp and neutral filters and is controlled by a reference photometer. In diodes at low voltages (≤ 1 V), two maxima were observed in the photosensitivity spectrum. The maxima were in the wavelength regions of $\lambda \sim 0.82$ μm and $\lambda \sim 0.86$ μm (Figure 1.3, curves 1 and 2). With the increase in the illumination level, the maximum in the impurity region became dominant ($\lambda \sim 0.86$ μm, curve 1). At bias voltages of 1 V or more, the impurity peak was imperceptible against the background of the long-wave downturn.

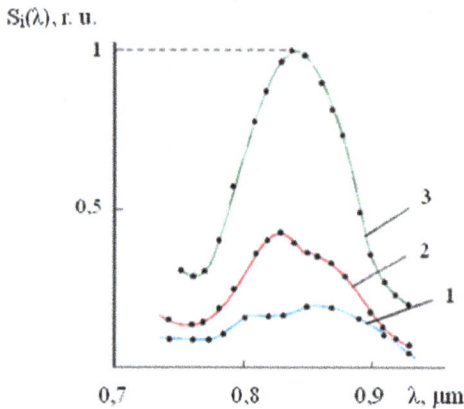

Figure 1.3: The spectrum of diode photosensitivity at forward bias $V = 0.5$ V and at different powers of incident radiation in W, μW: 1.3 (curve 1); 0.5 (curve 2); 0.1 (curve 3).

The photosensitivity spectrum of diodes did not reveal any distinctions at reverse bias. The maximum photosensitivity corresponded to the absorption band edge. Only a slight broadening of the maximum band was observed with the decrease in the power of the incident radiation.

1.1.4 Evaluation of photosensitivity and noise of diode structures

According to the current, the spectral photosensitivity $S_i(\lambda)$ of diodes is defined as the value of the photocurrent generated by the unit flux (or power) of radiation (Bosio, 2023; S. Khudaverdyan et al., 2005; S. K. Khudaverdyan, 2003; Sze et al., 2021; I. M. Vikulin et al., 2008):

$$S_i(\lambda) = \frac{I_L - I_D}{W} = \gamma \frac{\lambda}{1.24} \text{ A/W} \tag{1.3}$$

where $\gamma = N_e / N_{Phot}$ is the quantum efficiency which is defined as the ratio of the number of photoelectrons N_e passed through the diode to the number of absorbed photons N_{Fot} per time unit; λ is the wavelength of the incident radiation, microns;

$I_{\mathrm{Ph}} = I_{\mathrm{L}} - I_{\mathrm{D}} = qN_{\mathrm{e}}$ is the photocurrent defined as the difference between light and dark currents; $W = N_{\mathrm{Phot}} \times h\vartheta = N_{\mathrm{Phot}} \times hc/\lambda$ is the power of the radiation incident on the sample; c is the speed of light.

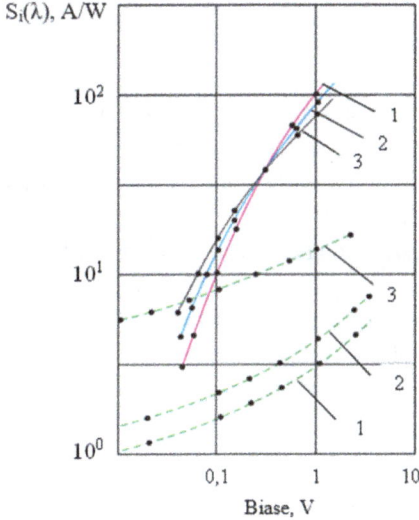

Figure 1.4: The dependence of the current photosensitivity $S_i(\lambda_\mathrm{m})$ of diodes on the bias voltage at different levels of illumination in W, μW: 1.3 (curve 1); 0.5 (curve 2); 0.1 (curve 3); ---- forward bias; —— reverse bias.

The diodes in the ultra-linearity region of the light CVC (Figure 1.2) are characterized by a decrease in photosensitivity as the power of the radiation falling on the sample increases (Figure 1.4, curves 1, 2, and 3). The photosensitivity weakly depends on the excitation level where the CVC passed on to the linear section and was about 100 A/W at the bias voltage of 10 V. For diodes, the quantum efficiency at 10 V was about 10^2.

The noise of the devices was measured at a frequency of 100 Hz in the 5 Hz band. The voltage noise of the measuring apparatus and the input resistance were 1 μW and 10^7 ohm, respectively. The measurement data showed that at bias voltages up to 10 V, the noise for diodes was lower than the noise of the apparatus ($\sim 1 \mu$V). With the increase in the bias voltage by about two times from the indicated values, the current noise coincided with the shot noise calculated by the formula (Sze et al., 2021; I. Vikulin & Stafeev, 1990):

$$\overline{I}_{\mathrm{N}}^{2} = 2qI\Delta f \tag{1.4}$$

where the frequency band was $\Delta f = 5$ Hz, and $I = I_{\mathrm{D}} + I_{\mathrm{Phot}}$. At low levels of illumination ($\sim 5 \mu$W/cm^2) and the bias voltages mentioned above, the threshold sensitivity calculated by the formula $F_{\mathrm{thr}} = \sqrt{\overline{I}_{N}^{2}/S_i(\lambda_\mathrm{m})}$ was lower than 10^{-14} W $\times$ Hz$^{-1/2}$ for all types of diodes, and the detection ability calculated by the formula $D^* = \sqrt{S}/F_{\mathrm{thr}}$ (the photosensitive area of diodes $S = 10^{-2}$ cm^2) was more than 10^{13} Hz$^{1/2}$/W $\times$ cm. It should be noted that the photoresistors based on the initial material of the diode had the threshold sensitivity of about 10^{-13} W × Hz$^{-1/2}$ at $V = 10$ V.

1.1.5 Speed of diodes

To measure the speed of the diodes, rectangular pulses of monochromatic light were applied to the photosensitive surface. Their duration allowed the photocurrent to reach the maximum value generated by continuous illumination. Before the measurements, the diodes were kept in the dark for a long time. The speed performance was assessed with the help of an electromechanical modulator.

The experimental equipment made it possible to change the frequency of light pulses over a wide range. The criterion for evaluating the inertia of photodiodes was the pulse rise t_1 and fall t_2 time constants. It was determined along the time axis of the oscilloscope, which corresponds to 0.63 parts of the change in signal level. The rise and fall time of the photodiode photocurrent was 10^{-5} s at 1.5 V and 1 μW.

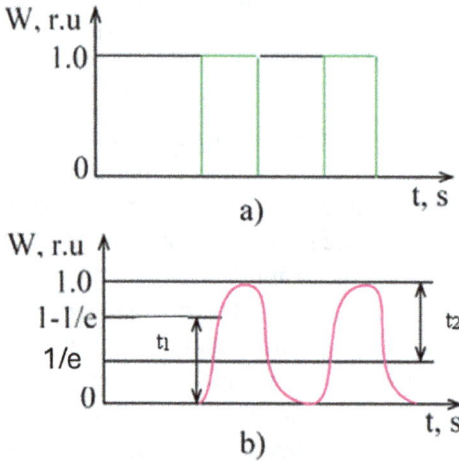

Figure 1.5: The appearance of the light signal pulse (a) and the appearance of the diode photocurrent pulse (b).

Figure 1.5 shows the time characteristics of the diode. The shape of the light pulse is shown in Figure 1.5a, and the shape of the photocurrent pulse in Figure 1.5b. The photocurrent sharply rises to a certain value and then slowly reaches a constant value. After turning off the light pulse, first, there occurs a sharp drop in the photocurrent, and then, a slow decrease to zero. The rise time t_1 and fall time t_2 in the range of bias voltage 1–2 V and the incident beam power of 10^{-7} and 10^{-6} W are approximately the same and occur at 5×10^{-5} s.

Figure 1.6 shows the dependence of the photocurrent of the diode on the frequency of the light signal at a constant direct bias voltage and different levels of illumination. It can be seen that the speed of the diode decreases as the illumination level decreases.

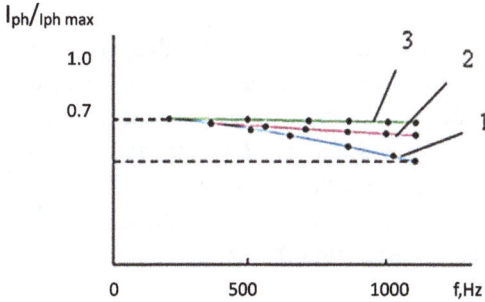

Figure 1.6: The dependence of the photocurrent of the diodes on the frequency of the light signal at different levels of illumination in μW: 0.5 (curve 1); 1.3 (curve 2); 2 (curve 3). Forward bias: $V = 2$ V.

This is explained by the fact that the available levels of attachment cause slow absorption of the photocurrent and play an important role at low illumination levels when the density of light-generated charge carriers is proportional to the density of unoccupied attachment centers.

1.1.6 Mechanism of photosensitivity of diode structures by energy diagrams

To explain the photosensitivity mechanism, the following studies were carried out and the energy band diagrams were studied. It is known that as a result of thermal treatment, CdTe single crystals change their resistivity (Lingg et al., 2018; Ojo & Dharmadasa, 2019). The evaluation of the latter is necessary for the determination of the position of the Fermi level. To make diodes, the n-type CdTe virgin single crystals with a specific dark resistance of $\rho_n = 10^7 - 10^8$ ohm $\times$ cm were used. The short-term thermal diffusion $(T = 823$ K, $t = 1$ min) did not change the resistivity of the initial material.

The density of free electrons in the conduction band of the initial material with the electron mobility of $\mu_n = 600$ cm^2/V $\times$ s was

$$n_n = \frac{1}{\rho_n \mu_n q} \approx 10^8 - 10^9 \text{ cm}^{-3} \tag{1.5}$$

and the Fermi level was $\Delta F_n = kT \ln(N_c/n_n) = 0.5 - 0.6$ eV lower than the bottom of the conduction band. At the illumination of the diodes from the side of the semitransparent silver contact, the maximum value of the photo-emf was 0.3 eV with the power of the incident light from the intrinsic absorption region up to 1.5 μW. It means that the height of the potential barrier φ_0 formed after the thermal diffusion of the junction was not less than 0.3 eV. The sign of the photo-emf shows that it is mainly determined by the barrier of the p-n junction formed between the high-resistance layer and the initial material. Since $\Delta F_n + \varphi_0 > E_g/2$, it can be concluded that a high-resistance layer with the p-type conductivity was formed near the single-crystal boundary through which the thermal diffusion of silver took place.

The diffusion depth of silver into the virgin single crystal n-CdTe with the resistivity of $\rho_n \sim 10^7 - 10^8$ ohm $\times$ cm was also evaluated by the angle-lapping method. The diodes that are $\sim 800\,\mu$m thick were made at the silver annealing temperature of 823 K and the annealing time $t = 10$ s. The depth the silver diffused into was $l_p = 4\,\mu$m, and the diffusion coefficient was $D_{Ag} \sim 4 \times 10^{-11}$ cm^2/s. The latter is slightly different from the diffusion coefficient of gold (Ojo & Dharmadasa, 2019). The penetration depth of silver into the CdTe single crystal at temperature and diffusion time $T = 823$ K and $t = 1$ min (Sze et al., 2021):

$$l_{Ag} = 2\sqrt{Dt} \approx 1\ \mu m \tag{1.6}$$

Taking the thickness of the p-region as $\sim 1\,\mu$m when the overall thickness of the diode is 800 μm, it is possible to determine the resistivity of this region. It makes 10^{10} ohm $\times$ cm, which corresponds to the carrier density of $p_n \sim 10^7$ and the position of Fermi level from the valence-band edge $\Delta F_p \approx 0.73$.

As it was noted before, in most cases, the virgin single crystal (the diodes were made from) did not change its resistivity and photoelectric properties during the silver firing for 1 min at $T = 823$ K. It was tested on samples that passed the same thermal treatment process, and then the contacts from indium were sprayed on them from two opposite sides. To form ohmic contacts, the In-n-CdTe-In structures were held at a temperature of 473 K for 5–10 min in an argon atmosphere. It is known that significant changes in the resistivity (to the point of overcompensation) and the photoelectric properties of n-CdTe single crystals occur during long-term vacuum annealing or at low pressure of cadmium vapor at sufficiently high temperatures $T = 773 - 873$ K. The evaluations show that in the process of short-term silver firing, carried out in our conditions, only a thin high-resistance layer with p-type conductivity is formed near the surface where the silver film is applied. The thickness of the layer is about 1 μm.

It should be noted that the evaluation of the thickness of the high-resistance layer agrees with the experimental fact stating that the maximum photo-emf appears when the sample is illuminated by no longer big powers of the light $(W = 10^{-6}$ W), incident from the side of the surface, and its further increase does not lead to an increase of the photo-emf. It means that the distance between the p-n junction and the surface is not more than the reciprocal of the absorption coefficient. According to (Medevedev, 1970; Kikoin, 1976), $1/a < 2\,\mu$m.

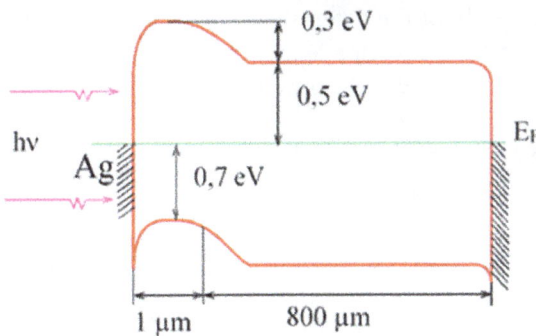

Figure 1.7: Energy diagram of diodes.

Figure 1.7 presents the energy diagram of diodes built based on evaluations. It is assumed that the space charge region in the contact layer with silver is restricted by the concentration of electrically active centers $N_a \geq 10^{14}$ cm^{-3} (Medevedev, 1970) and is $\leq 0.5\ \mu m$.

1.1.7 Mechanism of photosensitivity of diodes at forward bias

It is known that CdTe, based on the growth process and thermal annealing, with one of the components in excess, contains multiply charged acceptor centers that bond both with the complex including the intrinsic defect (Medevedev, 1970) and the charge states of cadmium vacancy (Grigoriev & Meïlikhov, 1997; Kikoin, 1976; Ojo & Dharmadasa, 2019). The increased photosensitivity of materials is explained by the presence of r-centers of slow and S-centers of fast recombination. The capture cross section of free electrons on S-centers, as a result of their acceptor nature, is much larger than the capture cross section on r-r centers. When illuminated from the intrinsic absorption region, the charge exchange of local centers takes place. S-S centers are almost completely filled up with electrons, and the recombination takes place through r-centers. When the concentration of r-centers is much larger than the concentration of S-centers, the concentration and the lifetime of free electrons and, consequently, the photosensitivity of the materials increase. This mechanism explains the photosensitivity of the photoresistor among the check samples made from the initial material of diodes.

However, the presence of nonlinearity in the CVCs in the dark and under illumination can be fully explained by the theory of a long diode. Indeed, the semilogarithmic CVCs of the diode in the case of direct bias (Figure 1.8) in the voltage range 0.3–1.8 V is approximated by an exponential dependence (see expression (1.1)). At high voltages, the CVCs increase with a near-linear dependence. The exponent c and the current I_c (see expression (1.1)) are determined, respectively, by the slope of the exponential site of the direct branch of CVCs (Figure 1.8). The ratio l/L was determined

(I. Vikulin & Stafeev, 1990) with the help of the determined constant C, the value $b = \mu_n/\mu_p \approx 10$ for CdTe, and relation (1.2). Then the length of the diffusion bias L was found. The lifetime of minority carriers τ was determined from (1.2).

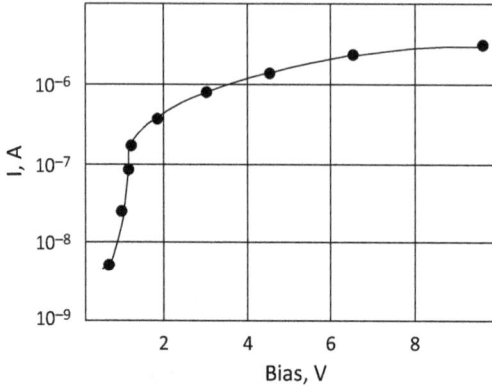

Figure 1.8: Direct branch of current-voltage characteristics. Incident beam power in μW, $\lambda \sim 0.8\,\mu m$.

The above-stated mechanism allows us to explain the ultra-linear increase of the current with the increase of the illumination intensity. Diodes differ in a smaller thickness of the high-resistance layer and a smaller ratio of the resistivity of the layer to the resistivity of the base region: $\rho_{np}/\rho_b = 10^2 - 10^3$. At that, for diodes, the ratio of the resistance of the layer to the resistance of the base region ρ_b in the dark is about one. Under illumination, the resistance of the thin layer of diodes is modulated to the full depth $(ad \leq 1)$, and the resistance of the base region becomes significantly greater than the resistance of the illuminated layer. At that, almost the entire voltage drops as a result of its redistribution falls on the base region. The further increase in the illumination intensity decreases the resistance of the layer but does not lead to a significant increase in the voltage drop across the p-n junction. For this reason, no ultra-linear dependence of the current on the illumination intensity is observed in diodes.

The observed regularities in spectral characteristics of diodes may be explained in the following way. The quantum energy corresponding to the long-wave edge of the short-wave maximum in diodes (Figure 1.3, curves 1 and 2) corresponds to the energy gap width of CdTe. The second maximum corresponds to the impurity absorption in CdTe. An increase in the fraction of impurity conductivity with an increase in the illumination level at low bias voltages is due to the high level of recombination of the intrinsic charge carriers near the illuminated surface, which, in turn, may be the reason for the short electron lifetime and the observed sublinearity on the lux-ampere characteristic (LAC). The probability of the participation of photoelectrons in electrical conduction increases with an increase in the bias voltage. Their lifetime increases too. At sufficiently high bias voltages, the number of recombination events of the carriers generated by light becomes negligibly small (due to the fast extraction of electrons into the contact by the field). The electron lifetime reaches its maximum, thus providing the LAC dependence which is close to linear and the prevalence of the intrinsic conductivity on the spectral characteristic.

As noted above, the positive electrode in the directly connected sample is located near the region of the maximum light absorption, thus providing the fast extraction of generated electrons into the contact. This reduces the trap occupation density at the bottom of the conduction band and may lead to a "decrease" in the intrinsic absorption energy by $\sim kT$ and to the shift of the shortwave maximum from the wavelength region of $\lambda \sim 0.8\,\mu m$ to the region of $\lambda \sim 0.84\,\mu m$. At reverse bias, the effect of the extraction of generated electrons into the contact is absent. The traps located at the distance of $\sim kT$ from the bottom of the conduction band are filled, and the maximum photosensitivity corresponds to junctions like the band-band. The conductivity at the expense of the impurity absorption observed at forward bias does not play a significant role, since the holes formed during the long-wave absorption of light at the depth from the surface generally have time to recombine on the way to the negative electrode.

1.2 Summary

The photoelectric properties of injection photodiode structures based on n-CdTe are described in this chapter. The existence of the injection amplification in these devices at forward bias and under illumination from the intrinsic absorption region is experimentally demonstrated. It is shown that the injection amplification is the result of the conduction modulation of the high-resistance layer and the base region and of the redistribution of the voltage drop across the diode structure. It is shown that to obtain the high photosensitivity of diode structures in the longitudinal illumination mode at forward bias, it is necessary not only to create a high-resistance photosensitive layer but also the optimum relationship between the resistivity, the layer thickness, and the base region. The basic parameters of the structures (the maximum spectral sensitivity and the detection ability D^*) that depend on the bias voltage are determined. The operating voltage is taken as 10 V based on n-CdTe, since above this voltage, the noise sharply increases. The operating bias voltage of diodes also depends on the choice of the above-mentioned relationships. To ensure the maximum detection ability, the operating voltage applied in the dark should almost completely drop across the high-resistivity regions of the structure. The p-n junction voltage is small, the injection is small, and therefore, the dark currents and noise are also small. But when illuminated, the operating voltage rate must be sufficient to ensure the injection through the p-n junction and the additional injection modulation.

Chapter 2
Cadmium telluride-based highly sensitive photosensors

2.1 Introduction

Cadmium telluride (CdTe)-based highly sensitive photosensors represent a cutting-edge advancement in optoelectronic technology. These photosensors are revolutionizing various fields, including medical imaging, security systems, astronomy, and environmental monitoring, due to their exceptional sensitivity, efficiency, and versatility. CdTe is a compound semiconductor composed of cadmium and tellurium, exhibiting unique properties that make it an ideal material for photosensing applications. The existing semiconductor detectors operate mainly under the reverse bias of the surface barrier and the diffusion p-n junctions or as homogeneous detectors (like photoresistors), thus presenting certain limitations to sensitivity. The process of injection amplification is highlighted in this chapter, causing the sensitivity of such detectors to record even the background radiation of the Earth.

2.2 Cadmium telluride-based injection detectors for ionizing radiation

The principle of the creation of photodiode structures with high sensitivity to visible radiation, based on the injection amplification (I. M. Vikulin et al., 2008; I. Vikulin & Stafeev, 1990), may be realized in the X-ray region of the spectrum. It is known that, in CdTe, the depth of the effective radiation absorption from the soft X-ray region does not exceed 100 μm up to the quantum energy of 40 – 50 keV (S. K. Khudaverdyan, 2003), and the intrinsic light is absorbed at the depth of ~2 μm (S. K. Khudaverdyan et al., 2014; Ojo & Dharmadasa, 2019) from the surface. Consequently, the optimum size of the detector (the length of the diode structure and the relation of the thicknesses of its different regions) may be somewhat different, depending on the energy of the registered quanta. However, the registration principle, which is based on the conductivity modulation and the injection amplification, will basically remain the same, with some distinctive features that may be attributed to the following:

1. Due to the higher penetrability of the X-rays, the electron-hole pairs may be generated not only in front of the high-resistance layer and the p-n junction region but also in the base region. Therefore, to effectively register the deeply penetrated radiation, it is necessary to create a sufficiently "thick" high-resistance layer, or, if it is technologically difficult to realize, to choose a high-resistivity base region. The conductivity modulation of the base region occurs behind the p-n junction,

https://doi.org/10.1515/9783111428024-003

which remained in the dark when illuminated with its light, and its conductivity modulation took place as a result of the voltage redistribution and the injection of minority carriers.

2. The generated charge carriers are "hot." Thus, the probability of Auger recombination increases. In the longitudinal irradiation mode, the change of the recombination mechanism is significant for the distribution of the nonequilibrium carriers along the sample length. The conditions of the injection amplification can strongly affect the output characteristics of the diode structure.

3. In conventional diodes with the p-n junction, running under reverse bias, the operating length of the detector by an order of magnitude is equal to $L = L_n + d_{p-n} + L_p$, where L_n, L_p are the diffusion lengths of minority carriers and d_{p-n} is the width of the p-n junction, while in diode structures running under forward bias, in the injection amplification mode, this length will be more. This is attributed to the fact that the regions beyond $L_n + d_{p-n} + L_p$ play an active role in it, which is further increased if there occurs the generation of nonequilibrium carriers by the X-radiation quanta.

2.3 Mechanism of free carrier generation upon irradiation by high energy photons

When the diode structure is irradiated by high energy quanta, photo- and Compton electrons appear in the bulk of the semiconductor. The Compton effect appears when the quantum energies are about 10 keV. The flux density of these secondary electrons with the depth X increases from zero near the surface up to the maximum value at the depth equal to the range of the electron X_e with the energy close to the energy of the incident quantum. When $X \succ X_e$, the flux density of secondary electrons and the flux density of quanta (Kikoin, 1976; Moiseev & Ivanov, 1984) begin to decrease with depth, that is, according to the law $I_x = I_0 \exp(-\chi X)$, where χ is the attenuation coefficient.

The generated electrons spend their energy on ionization, excitation of the atoms of the semiconductor, and bremsstrahlung. Some part of the energy, which is small as compared with the ionization losses, is transferred to the atoms of the substance, thus inducing radiation defects. As a result, some quantity of nonequilibrium charge carriers – electrons and holes, is generated in the bulk of the diode. The concentration of these carriers depends on the radiation intensity and lifetime. For homogeneous semiconductors, knowing the average energy w required for the generation of one pair of carriers, which is the ratio of the total ionization loss to the number of generated pairs of nonequilibrium carriers, it is possible to find the energy E_n absorbed in the bulk of the semiconductor by the number of experimentally measured pairs of carriers:

$$E_n = wN \tag{2.1}$$

For CdTe, $w = 4.65$ eV (Ivanov, 1988).

When irradiated to determine the absorbed energy of diode structures in the injection amplification mode, it is necessary to know the injection amplification rate, which can be defined in the following way:

$$Y = \frac{S_i}{S_{iR}} \langle V_r = V \rangle = \frac{(\partial I / \partial W)}{\partial I_R / \partial W_{V_R}} \langle V_r = V \rangle \qquad (2.2)$$

where S_i is the differential dose sensitivity of the diode, S_{iR} is the differential dose sensitivity of the "conjugate detector," that is, detector of the same configuration as the diode, from the same source material, but with noninjecting contacts, and W is the dose rate (or radiation intensity).

The object of research is diode structures, as shown in Figure 2.1, based on n-CdTe with a thickness of ~800 μm and $\rho \sim 10^7 - 10^8$ ohm × cm. The high-resistance p-type region with a thickness of ~3.5 μm is created by thermal diffusion of silver at $T = 850$ K in the argon atmosphere for 1.5 min.

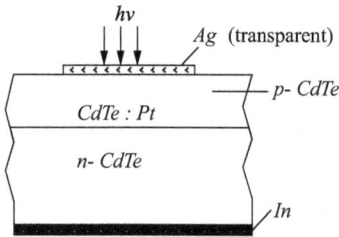

Figure 2.1: Cadmium telluride-based detector structure.

The n⁺-type rear contact to diode structures is created in the process of annealing the film of indium. To measure the parameters of the obtained structures, the facility used consists of X-ray generators such as molybdenum, aluminum, copper, steel tubes, electrometer TR-1501, and a DC voltage source.

2.4 Diode characteristics and parameters under X-ray irradiation

Below are the results of the studies of the properties of diode structures exposed to X-radiation (molybdenum). The measurement of the X-radiation dose was made by the dosimeter, which was placed at the same distance from the source as the diodes when their properties were studied. It allowed us to ignore the loss of X-ray quanta in the air gap between the source and the receiving surface of diodes. When measuring, the silver contact was believed to completely transmit the X-ray quanta (the film thickness Ag $\leq 0.1\,\mu$m), since to halve the X-ray flux, the silver layer with the thickness of 23 μm was required (S. K. Khudaverdyan & Khachatryan, 2019; Mirkin, 2012). The studies were carried out in the current operation mode.

Figure 2.2 shows the dependence of the ratio of photocurrent and short-circuit currents I_{Ph}/I_{sc} on the bias voltage. The initial sharp rise indicates the existence of deep recombination centers in the space charge region (I. Vikulin & Stafeev, 1990). The photocarrier losses, caused by these centers, decrease with an increase in the bias. The voltage ranges from 1 to 6 (depending on the dose rate), which corresponds to the photodiode operation mode, and after 6, when the linear increase of the photocurrent begins on the voltage-current characteristics, it corresponds to the photoresistive operation mode.

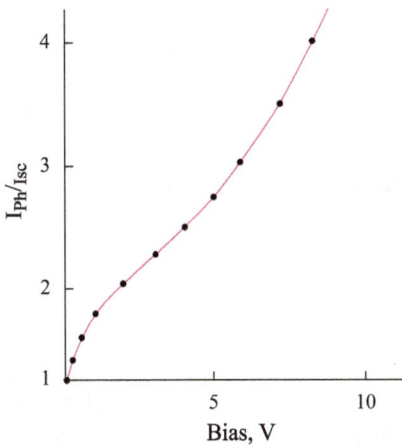

Figure 2.2: Relative increase of the photocurrent with the increase of bias voltage for diodes under reverse bias $(P \sim 5.1 \times 10^{-2}\ R/s)$.

Starting from the voltage $V = 0.5\,V$, the Voltage-Ampere Characteristics (VAC) of straight-biased diodes is superlinear when the sample is irradiated with X-rays. As the voltage increases, it gradually becomes linear.

The hyperlinearity index n at the beginning of the section as well as the bias voltage V at which the linearity of the CVC begins increase with the increasing beam power, and when the power is $5.1 \times 10^{-2}\ R/s$, $n \sim 3$ and $V \approx 10\,V$. When the sample is irradiated from its own absorption region, the linearity begins earlier ($V \approx 5\,V$), and the changes with the change in the illumination power are not as noticeable as those that occur with the change in the power of the X-ray beam.

The analysis of the current-voltage characteristics in the semilogarithmic scale (Figure 2.3) revealed the initial exponential section corresponding to the range of the current changes by an order of magnitude and more, depending on the dose power (Figure 2.3, curves 1 and 2).

The calculated values of the constant c from eq. (1.1), describing the exponential law, are 16 and 12 for the powers $1.1 \times 10^{-4}\ R/s$ and $5.1 \times 10^{-2}\ R/s$, respectively, and the corresponding lifetimes of minority carriers turn to be equal to $8 \times 10^{-5}\ s$ and $10^{-4}\ s$. That is, similar to the mechanism of the operation of diodes, when illuminated by its own light, the conductivity modulation of the n-base is conducted by an additional

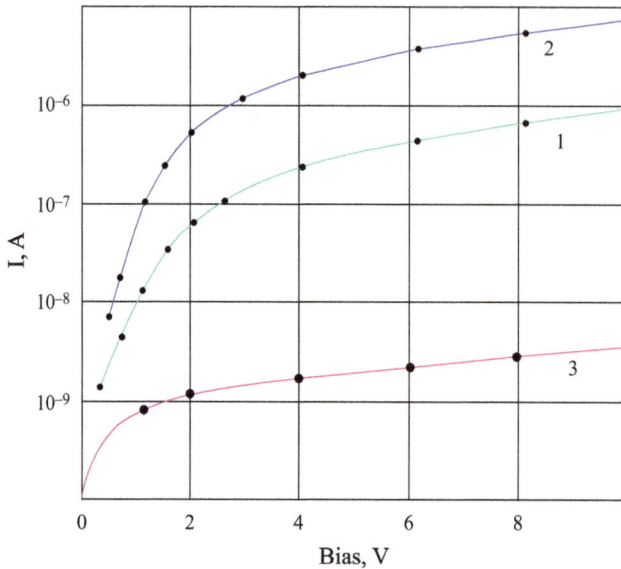

Figure 2.3: The direct branch of the diode at the power dose (R/s): (1) 1.1×10^{-4}; (2) 5.1×10^{-2}; and (3) 0.

c-mechanism. It should be noted that taking ratios into account the relationship between the quantum energy of X-radiation (17 keV) and intrinsic radiation (1.5 eV), and the energy of electron-hole pair generation (4.5 eV) (S. K. Khudaverdyan & Khachatryan, 2019), the quantum efficiency (2.3) under X-radiation at $R = 1.8 \times 10^{-4}$ R/s turns to be equal to $\sim 3 \times 10^6$, which is considerably more than under the illumination by its own light.

These results can be explained if it is assumed that, as opposed to the radiation from the self-absorption region when the light modulates only the conductivity of the interlayer, the majority of X-ray quanta are absorbed in the depth of the semiconductor (in the n-region). This brings to the reduction of the resistance of the part of the n-base, and the additional redistribution of the voltage between the p-n junction and the high-resistivity p- and n-regions.

As a result, the amplification possibilities in the device increase due to the more intensive realization of the positive feedback, especially at low dose rates, since at high dose rates there is a probability of the increase of the absorption of X-ray quanta near the contact opposite the surface. Consequently, there is a probability of an increase in the losses of secondary electrons and a decrease in the registration efficiency.

2.5 Dose sensitivity of diodes exposed to *X*-radiation

The spectral dose sensitivity of the detector is determined by the relation

$$S_{iv} = \frac{I}{P} \tag{2.3}$$

where I is the current at the output of the detector and P is the dose rate measured at the disposition of the detector.

The dose sensitivity S_{iv} of diodes was decreased both under forward and reverse biases. This behavior of the dose sensitivity is due to the sublinear dependence of the current on the dose. The dose sensitivity of diodes was increased both under forward and reverse biases with the increase of the voltage. In the voltage range from 0.1 to 20 V, applied to the diode, the dose sensitivity changed by $2-3$ orders of magnitude and reached the values of 10^2 A/R × s under forward bias. Under reverse bias, the dose sensitivity was by 1.5–2 orders of magnitude lower than under forward bias ($V = 10$ V). For diodes, the dependence of the quantum efficiency on the voltage under forward bias (Figure 2.4) is calculated by the formula

$$\Upsilon = \frac{I - I_T}{q \times N} \tag{2.4}$$

where I is the current at the X-radiation of the diode, I_T is the dark current, q is the electron charge, and N is the number of electron-hole pairs generated by the X-radiation per time unit.

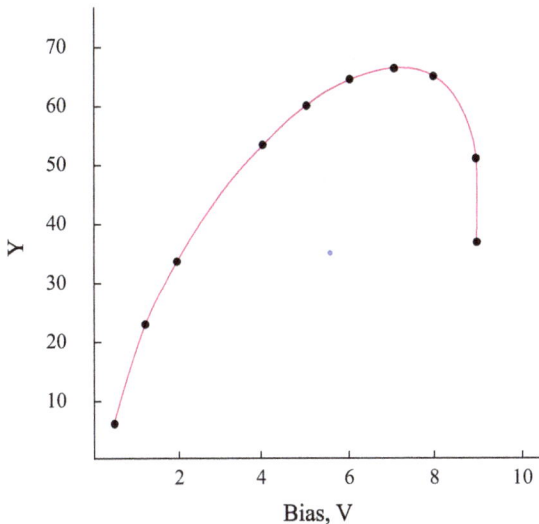

Figure 2.4: Dependence of injection amplification rate on bias voltage.

The ultra-linear growth of the quantum efficiency is observed with the increase of the voltage. At high bias voltages, the ultra-linear dependence $\Upsilon(V)$ turns into linear dependence. In this voltage range $(V \geq 10\,\text{V})$, the current flowing through the diode is limited by the modulated resistance of the base region.

The coefficient of the photoresistive amplification is determined by the expression (I. Vikulin & Stafeev, 1990)

$$G \approx \frac{\tau}{t_{for}} = \frac{\mu_n V \tau_n}{l^2} \tag{2.5}$$

where τ_n is the lifetime of the majority carriers, t_{for} is the transit time of free carriers through the base region, and l is the transit length (the thickness of the sample). When the voltage $V = 10\,\text{V}$, the lifetime of electrons $\tau_n \sim 10^{-5}$ s (I. Vikulin, Kurmashev, & Stafeev, 2008), and the mobility $\mu_n = 600\,\text{cm}^2/(\text{V}\times\text{s})$, $G \approx 10$, for diodes.

In the case of the photoresistive effect at $\mu_n = 600\,\text{cm}^2/(\text{V}\times\text{s})$ and the dose rate $6.5\times10^{-2}\,\text{R/s}$, when 1×10^9 pairs of carriers are generated in the homogeneous CdTe, the current flowing through the photoresistor with the thickness 200 μm and the voltage 10 V will be equal to (I. Vikulin & Stafeev, 1990):

$$I = qn\mu ES \sim 3\times10^{-6}\,\text{A} \tag{2.6}$$

The injection amplification coefficient (Figure 2.4) is calculated by the ratio (1.1). The linear characteristic is taken for the current-voltage characteristic of the conjugate photoresistor. The former coincides with the diode characteristic on the ohmic section at low voltages when the injection is still small. The dependence shows that the injection amplification grows with the increase of the applied voltage in the forward direction up to $V = 10\,\text{V}$, at which the maximum is observed. At the maximum, the current sensitivity is 65, which is higher than the current sensitivity of the conjugate photoresistor. The decrease of the injection amplification coefficient at $V > 7\,\text{V}$ can be explained by the current limited by the resistance of the base region and the resistance of the interlayer.

2.6 Thresholds and detection ability of diodes under X-radiation

The evaluation of the threshold sensitivity was carried out by analogy with the evaluation of the threshold sensitivity for photoreceivers with the help of formulas (Sze et al., 2021; I. Vikulin & Stafeev, 1990):

$$F_{thr} = \frac{\sqrt{I^2_{nois}}}{S_{iV}}, \quad D^* = \frac{\sqrt{S}}{F_{thr}} \tag{2.7}$$

Since the current noises within the bias voltage range of up to 15 V (+ on Ag electrode) were less than 10^{-12} A and 3×10^{-4} A/R/s, respectively, at low radiation doses $V = 10\,\text{V}$, the evaluated threshold sensitivities under forward bias were $10^{-4}\,\mu R/s$ and $10^{-3}\,\mu R/s$, respectively. The detection abilities were equal to 10^3 and $100\,cm \times s/\mu R$. For diodes, these values were $10^{-2}\,\mu R/s$ and $10\,cm \times s/\mu R$ under reverse bias.

The average air radioactivity in the layer near the Earth's surface is $4 \times 10^{-3}\,\mu R/s$, and $2/3 \times (\sim 10^{-3}\,\mu R/s)$ of which is Earth's background (Shklovskii, 1987). The maximum environmental radioactivity in which the man is allowed to work is $0.9\,\mu R/s$. The photo-response of forward-biased diodes at this dose is $3-5$ times more than the dark currents at $V = 10$ V. Thus, the developed samples can measure the doses that are much smaller than the doses threatening for human life. The limiting registration ability of the samples is commensurable with the Earth's background.

2.7 Interconnection of sensitivity in the intrinsic and X-ray regions of the spectrum

When identifying the interconnections of the operation mechanisms of diodes in the intrinsic and X-ray regions of the spectrum, it is necessary to consider the consequences that arise due to the absorption of quanta with different energies. The intrinsic light quantum generates, at best, one pair of charge carriers. The X-radiation quantum can create secondary X-radiation quanta with lower energy, as well as hot electrons (S. K. Khudaverdyan, 1999). Both spend their energy on the reionization and the lattice thermal vibrations. The latter, using ionizing radiation, produces the average energy of the electron-hole pair generation, which is three times bigger than the energy gap width. The reionization cycle continues until the last generated particle loses its ability to impact ionization. As a result of the exposure to ionizing radiation, one quantum with an energy of more than 5 keV generates 10^3 or more electron-hole pairs. The amplification rate, calculated by formula (2.1) at $V = 10$ V, is equal to ~ 100. Considering that $w = 4.65$ eV, the quantum efficiencies under the illumination from the self-absorption region and under the low-intensity X-radiation $(P \sim 1.8 \times 10^{-5}\ R/s)$ are $\sim 10^2$ and $\sim 10^3$ in the same electric field.

Thus, in diodes, amplification occurs both in the light and under X-radiation. In the latter case, it significantly increases with a decrease in the radiation intensity. This difference in the sensitivity of diodes at their own light under X-radiation is due to the absorption of X-ray quanta, mainly in the depth of n-based within the range of $50\,\mu m$ from the contact region. This brings to the redistribution of the voltage between the remained high-resistivity n-region and the contact (also high-resistivity) p-region, thus increasing the voltage drop in the contact region and, consequently, the injection from the contact.

At the illumination from the intrinsic absorption region, the quanta are generally absorbed in the contact p-region, hereby reducing its resistance and decreasing the voltage drop on it, which leads to the decrease in the injection fraction. With the increase in the X-radiation intensity, there occurs an increase in the number of absorbed quanta in the contact region. As in the case with the own light, this can lead to the reduction of the injection. Simultaneously, the range of the secondary electrons increases, and the effect of the conductivity modulation of the n-base decreases. The number of centers responsible for the decrease in the lifetime of the charge carriers also increases. All that leads to a decrease in the sensitivity of devices to X-radiation.

2.8 Summary

The developed diode structures are based on CdTe and possess high sensitivity in the X-ray region of the spectrum. The properties of diode structures under X-ray exposure in the current operation mode have been studied. The injection amplification of the current under the influence of irradiation under forward bias has been experimentally proved. Under X-ray irradiation, the base region is modulated (since the penetration depth of X-rays in the CdTe is $10–50\,\mu m$ at quantum energies up to $20\,keV$), and the p-n junction is disposed at the distance of several microns from the contact exposed to radiation. That is why, for efficient injection amplification, the base region in X-ray sensors must be sufficiently high resistive. This makes the essential difference between the X-ray sensors and photodiodes. In photodiode structures, the base region remains "nonexposed," and its resistivity must be small so that the decrease in the resistance of the interlayer under illumination as a result of the redistribution of the voltage in the structure leads to an increase in the voltage drop on the p-n junction and, eventually, to the additional injection modulation of the resistance of the interlayer. The next chapter considers the functional potentiality of structures with a high-resistance thin layer between two oppositely directed potential barriers and the photoelectric and electrophysical processes in them. CdTe and silicon are used as initial materials for diode structures.

Some of the possible applications of these photodetectors include: (1) medical imaging: CdTe-based photosensors are used in X-ray detectors and computed tomography scanners for high-resolution medical imaging with reduced radiation exposure. They offer superior image quality and sensitivity, facilitating accurate diagnosis of diseases and abnormalities. (2) Security and surveillance: CdTe photosensors are employed in surveillance cameras, night vision systems, and motion detectors for perimeter security and surveillance applications. Their high sensitivity and low-light performance enable reliable detection and identification of intruders or suspicious activities. (3) Astronomy and space exploration: CdTe-based photosensors are utilized in telescopes, space probes, and satellite imaging systems for capturing celestial objects and studying cosmic phenomena. Their sensitivity to faint light enables astronomers to observe distant stars, galaxies, and celestial events with unprecedented clarity. (4) Environmental monitoring: CdTe photosensors are integrated into environmental monitoring devices for measuring air and water quality, detecting pollutants, and monitoring radiation levels. Their sensitivity and reliability make them valuable tools for environmental research and pollution control efforts.

Hence, CdTe-based highly sensitive photosensors represent a significant advancement in optoelectronic technology, offering exceptional sensitivity, efficiency, and versatility across various applications. With ongoing research and development, these sensors continue to drive innovation in fields ranging from healthcare to space exploration, promising exciting opportunities for future advancements in sensing and imaging technology.

Chapter 3
Cadmium telluride and silicon-based photodiodes

3.1 Introduction

To study the functionality of structures with a high-resistivity layer, the photoelectric and electrophysical processes occurring in them need to be considered. Cadmium telluride (CdTe) and silicon were used as source materials for diode structures. An interesting mechanism for changing the sign of the spectral photocurrent is discovered and discussed. The influence of bias voltage and radiation intensity on the spectral characteristics as well as the possibility of measuring wavelength are considered. Under conditions of oppositely directed potential barriers, the behavior of the current-voltage characteristics (CVCs) of photodetectors based on CdTe and Si, and their similarities and differences are discussed.

3.2 Cadmium telluride-based diodes

CdTe-based diodes with a high-resistance layer CdTe:Pt were fabricated from relatively low-resistance virgin n-type crystals with a dark resistivity $\rho \sim 10^2$ ohm $\times$ cm by 5-min thermal diffusion of platinum at a temperature of 823 °C under argon atmosphere. A semitransparent silver layer served as a contact from the side of the high-resistance layer and an aluminum layer served as a contact from the opposite side. The diodes, thus obtained, had the structure Ag-CdTe:Pt-n-CdTe-Al (Figure 3.1) and, at the total length of 400 μm, the high-resistance layer was ~2 μm (S. Khudaverdyan et al., 2005). When measuring the short-circuit current of the diode, the illumination was carried out through a semi-transparent Ag contact.

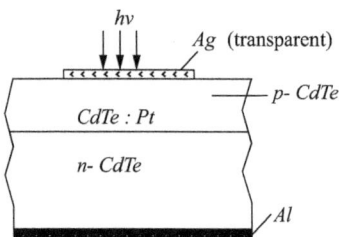

Figure 3.1: Cross section of the CdTe-based photodetector structure.

Figure 3.2 presents the experimental curves of the dependence of the short-circuit photocurrent on the wavelength for diode structures. As can be seen from the figure, there occurs a change in the spectral photocurrent sign.

https://doi.org/10.1515/9783111428024-004

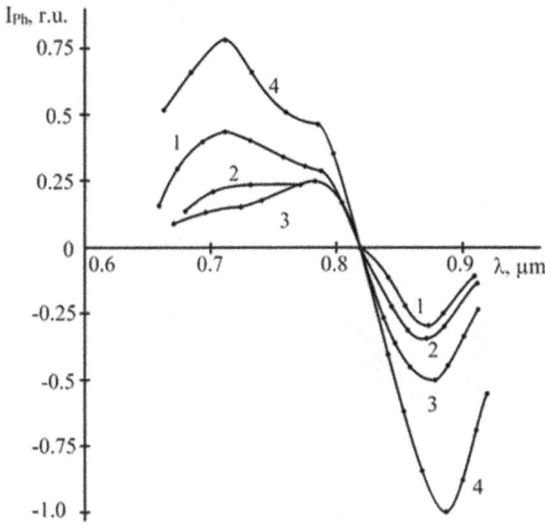

Figure 3.2: The dependence of the short-circuit photocurrent on the wavelength. The radiation power incident on the sample increases with an increase in the curve number.

The change of the photocurrent (S. K. Khudaverdyan, 1999, 2003) sign occurs when the inversion point ($\lambda_i \sim 0.82\,\mu m$) corresponds to the intrinsic absorption region of CdTe. The photocurrent increases with the increase of the intensity of the incident light both in the shortwave (to the left of the inversion point) and the longwave regions (to the right of the inversion point) of the spectrum. In the shortwave region of the spectral dependence of the short-circuit photocurrent, two peaks were observed at $\lambda = 0.71\,\mu m$ and $\lambda = 0.78\,\mu m$, which corresponded to the quantum energy of 1.74 and 1.59 eV, respectively, with the peak at $\lambda_1 \approx 0.78\,\mu m$ dominating at low light intensities (Figure 3.3, curves 1 and 2). With the increase of the light intensity, the maximum value of the photocurrent at $\lambda_2 \approx 0.71\,\mu m$ increased, and at the light power of $W > 0.5\,\mu W$, it was higher than at $\lambda_1 \approx 0.78\,\mu m$. Figure 3.3 presents the spectral dependences of the photocurrent at constant light intensity but at different bias voltages (positive potential on the Ag contact).

Two peculiarities were observed. Firstly, with the increase of bias voltage, the wavelength range of positive photosensitivity significantly increases. Secondly, the point of the change of the photocurrent sign moves toward the longwaves (at reverse bias, it moves toward the shortwaves).

3.3 Silicon-based diodes

The silicon-based diodes were made by recrystallizing a 1-μm-thick polycrystalline film n-Si ($n \leq 1.5 \times 10^{15}$ cm^{-3}) deposited on a titanium film. During the recrystallization,

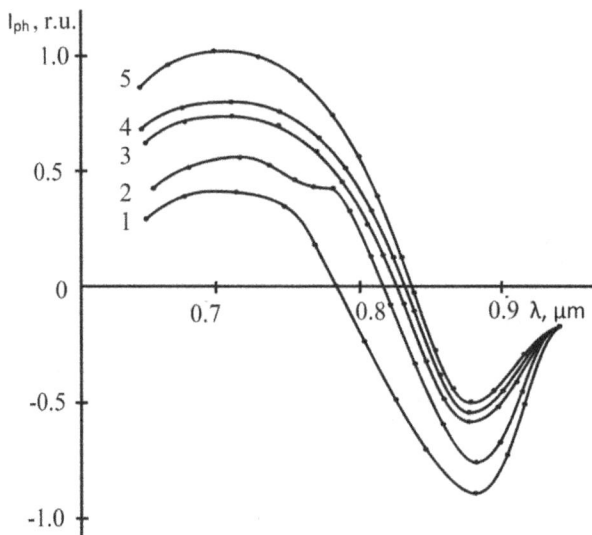

Figure 3.3: The spectral dependence of the photocurrent at a constant power of incident radiation, but at different bias voltages: "–" on the Ag contact; 1, 0.04 V; "+" on the Ag contact; 3, 0.02 V; 4, 0.04 V; 5, 0.06 V; 2, 0 V.

titanium silicide ($TiSi_2$, formation temperature was 600 °C) and a potential Schottky barrier $TiSi_2$-Si ~ 0.6 eV were simultaneously formed. The estimation of the barrier height was carried out according to the CVCs. The second Schottky barrier was a semi-transparent nickel layer deposited on a silicon film, which, too, formed a nickel silicide (NiSi) together with silicon (formation temperature was 400 °C). The height of the potential barrier was 0.5 eV (S. K. Khudaverdyan & Kocharyan, 2004; S. Kh. Khuda-verdyan et al., 2005). The illumination was carried out through semi-transparent nickel contact. The structure of the obtained NiSi-n-Si-$TiSi_2$ samples is shown in Figure 3.4.

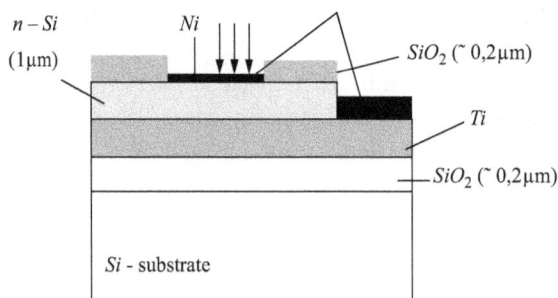

Figure 3.4: NiSi-n-Si-$TiSi_2$ structure.

The spectral characteristics of the photocurrent of silicon-based diode structures had certain peculiarities as the CdTe-based structures. Firstly, the change in the photocurrent sign was observed (Figure 3.5).

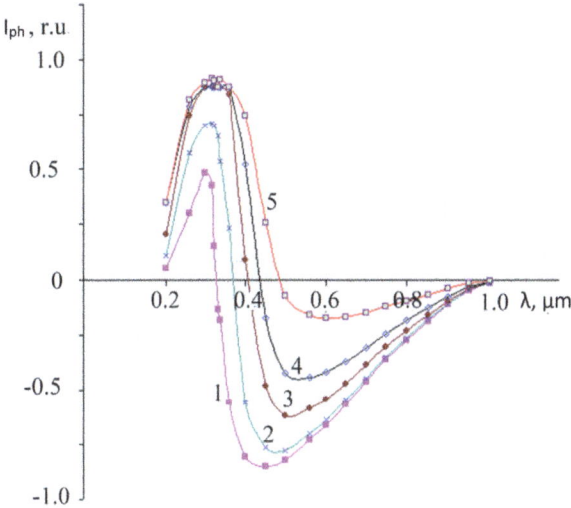

Figure 3.5: The spectral dependence of the photocurrent at different values of bias voltage in V: 0.36 (curve 1); 0.34 (curve 2); 0.27 (curve 3); 0.18 (curve 4); − 0.01 (curve 5).

Secondly, the position of the inversion point did not depend on the incident radiation power (Figure 3.7), but it depended on the external voltage (Figure 3.5) (S. K. Khudaverdyan & Kocharyan, 2004; S. Kh. Khudaverdyan et al., 2005). When increasing the external bias voltage "+" on Ni contact (Figure 3.7), the reduction of the p-type leg and the increase of the n-type leg of the spectral characteristics occurred both on the amplitude and the width (Figure 3.7). With the change of the polarity of the bias voltage, the inverse picture was observed.

The longwave photosensitivity did not reach the intrinsic absorption region, and the structure had mainly shortwave photosensitivity.

Figure 3.6 presents the dependence of the inversion point of the spectral photocurrent λ_{inv} on the bias voltage. The dependence is close to linear within the wavelength range of 0.35–0.5 μm. When the wavelengths are shorter than 0.35 μm, a sharp change of the value of λ_{inv} is observed even at a slight change in the bias voltage. Figure 3.7 shows the spectral dependence of the photocurrent at different radiation intensities. It is important to note that the point of change of the sign of the spectral photocurrent does not depend on the intensity of the recorded beam.

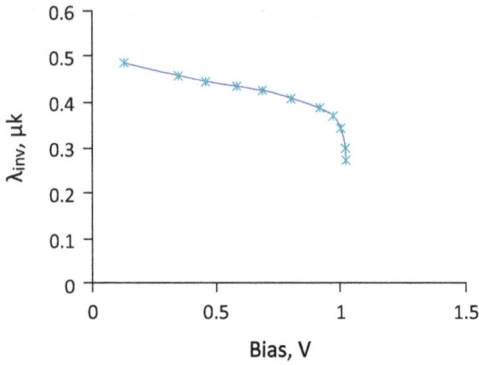

Figure 3.6: The dependence of the inversion point of the spectral photocurrent λ_{inv} on the bias voltage.

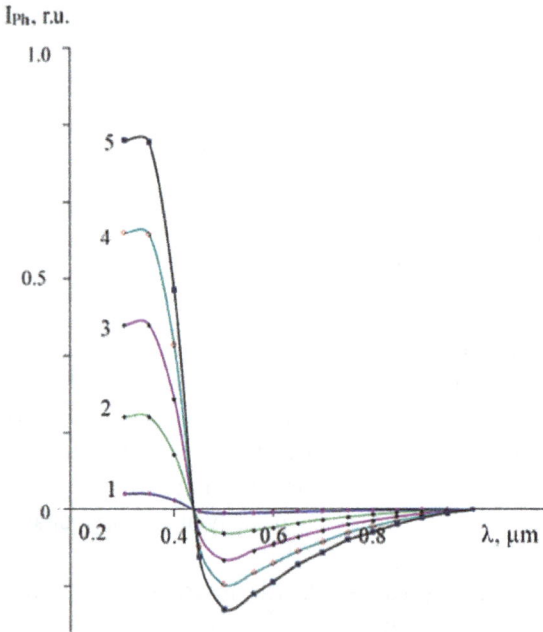

Figure 3.7: The spectral dependence of the photocurrent at $V = 0.18$ V and at different radiation intensities in quantum/m^2 s: 1, 10^{18}; 2, 6×10^{18}; 3, 1.2×10^{19}; 4, 1.8×10^{19}; 5, 2.4×10^{19}.

3.4 Mechanism of spectral photocurrent sign change in diode structures

To explain the photoelectric properties of diode structures, we will refer to the energy band diagrams of the structures in question. The energy band diagram of CdTe-based samples was built by taking into account the fact that the concentration of free electrons in the n-region was equal to $n_n \approx 10^{14}$ cm^{-3}. The calculations were carried out by eq. (3.1) at $\rho_n \sim 10^2$ ohm $\times$ cm. The Fermi level was 0.2 eV below the edge of the conduction band. The position of the Fermi level in the high-resistance layer was determined from the dark CVCs (Figure 3.8). It turned out that the Fermi level in the layer was not lower than 0.7 eV from the edge of the valence band.

To determine the width of the layer that depends on the depth of the platinum diffusion into the initial material, an oblique grid method was used. The diffusion depth of platinum determined by this method was 2 μm, and therefore the width of the layer was ~2 μm. The calculated diffusion coefficient of platinum $(D_{Pt} = 3.7 \times 10^{-11}$ cm^2/s) turned

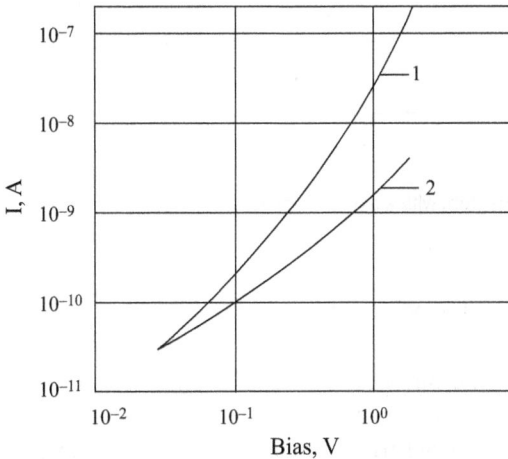

Figure 3.8: CVC of diodes in the dark: 1, "+" on the silver electrode; 2, "–" on the silver electrode.

out to be comparable with the data for the diffusion coefficient of gold $(D_{Au} = 3.5 \times 10^{-11}$ cm^2/s; Ivanov, 1988; Khudaverdyan & Khachatryan, 2019; Kikoin, 1976; Ojo & Dharmadasa, 2019).

The barrier height for holes due to the contact with silver was calculated by the difference between the work functions of silver and CdTe and made 0.4 eV. Figure 3.9 presents the band diagram of CdTe-based diodes. The diagram is built based on the results of the research. I_n, the high-resistance layer, has p-type conductivity, and its thickness is covered by the space charge region (SCR) of two barriers. As a result, the maximum potential energy is observed on the diagram.

The energy band diagram (Figure 3.9) of the structure based on CdTe (Figure 3.1) makes it possible to explain the results of photoelectric studies. At the "transverse" (lateral) illumination of the structure, a photo-emf occurs on the potential barriers of the p-n junction and the contact-metal junction in the open-circuit mode. Since the barriers are oppositely "switched on," the resulting photo-emf is less than the height

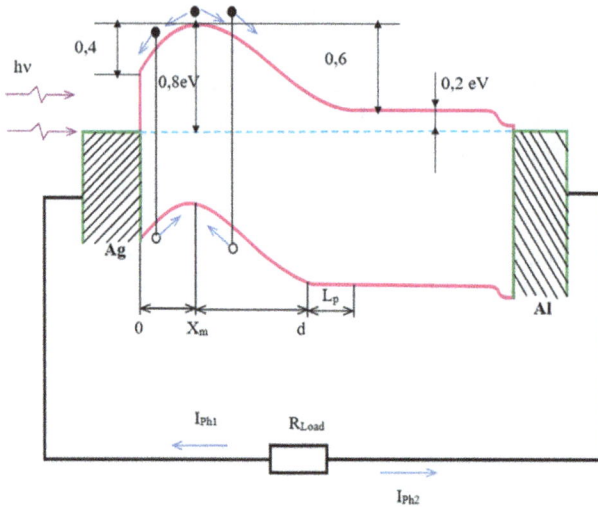

Figure 3.9: Short-circuit current in the diode structure with a high-resistance layer in the longitudinal illumination mode.

of each of the barriers. In the short-circuit mode, the current starts flowing in the external circuit.

The direction and the magnitude of the current are determined by the efficiency of the division of the electron-hole pairs on each barrier and by the ratio $(d - x_m + L_p)/x_m$, where $(d - x_m)$ is the thickness of the potential barrier from the side of the p-n junction, x_m is the thickness of the potential barrier from the side of the Schottky barrier, and L_p is the diffusion length of the holes in the low-resistance n-region.

At the illumination of the structure through the semitransparent metal (Ag) contact, the situation changes. If the photon energy $hv > E_g$, the radiation is absorbed near the surface, and the division of the electron-hole pairs occurs only in the space charge field of the contact region. The p-n junction region remains in the dark. The interlayer, with respect to the semitransparent contact, is positively charged, and, in the short-circuit mode, the current I_{ph1} starts flowing through the load resistance R_{Load} (Figure 3.10). With the attenuation of the energy of the incident photons, the light penetration depth increases. If the layer thickness is small, at the sufficient attenuation of the photon energy, the electron-hole pairs are generated in the p-n junction region, too, which results in the appearance of a photo-emf and a current I_{Ph2} of the opposite sign on the p-n junction, and the reduction of the current through the resistance R_{Load}.

Thus, when a greater number of photogenerated pairs is divided by the p-n junction, the ratio I_{Ph2}/I_{Ph1} can prove to be more than one. All this shows that at the illumination through the semi-transparent metal silver contact, in the short-circuit mode,

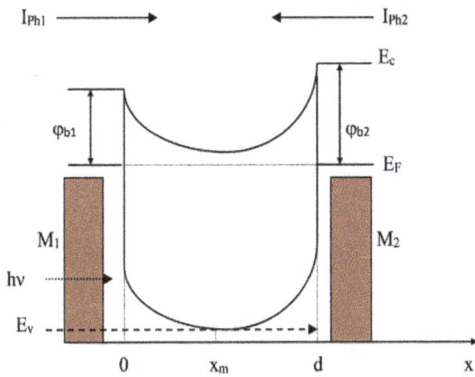

Figure 3.10: Energy band diagram of NiSi-n-Si-TiSi$_2$ structures. M_1 and M_2 are the metal contacts that create silicide potential Schottky barriers with heights φ_{b1} and φ_{b2}, respectively.

the change of the spectral photocurrent sign is possible with the attenuation of the photon energy.

If an upward bend of the bands takes place in the contact region of the SCR of the structure, the photocurrent has always one direction, and the change of the photocurrent sign is not observed. Unlike the CdTe-based samples, the silicon two-barrier structures were made by recrystallizing thin silicon films with the n-type conductivity $N_D = 1.5 \times 10^{15}$ cm^{-3} and the thickness $d = 1\,\mu$m. Two oppositely directed silicide barriers were formed on both sides of the film (Figure 3.10). At the selected impurity concentration, the width of the SCR of each barrier, calculated by the formula $l = \sqrt{(2\varepsilon\varepsilon_0\varphi_c)/q^2N_d}$, was more than half the width of the base $(d/2)$. As a result, the SCRs of the two barriers overlap (G. E. Grigoryan et al., 1997; S. Khudaverdyan et al., 2021; Khudaverdyan et al., 1998; S. K. Khudaverdyan, 2003; S. Kh. Khudaverdyan, 2003) and cover the entire thickness of the silicon film. On the energy band diagram of such a structure, the minimum of the potential energy of electrons is formed in the point x_m (Figure 3.11).

To substantiate the obtained experimental results, it was necessary to deduce an analytical expression of CVCs for the complete current consisting of the dark current and the current at illumination. For that purpose, the dependence of the SCR width on voltage was analyzed, and the expressions for the width of the SCRs of two-barrier diode structures with a high-resistance layer were obtained.

3.5 Dependence of the thickness of the energy barriers on the external voltage in structures with a high-resistance layer between Schottky barriers

After the example of silicon samples, we shall determine the potential distribution in the SCR of two-barrier structures (Figure 5.12). To do this, we solve the Poisson equa-

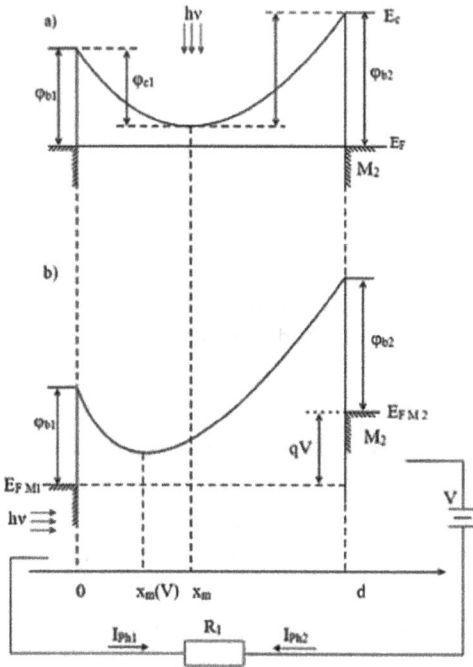

Figure 3.11: Energy band diagram of structures with two Schottky barriers: (a) in the absence of the external voltage and (b) in the presence of the external voltage (reproduced with permission).

tion connecting the field potential $V(x)$ with the volume density of the charges creating that field

$$\frac{d^2V(x)}{dx^2} = -\frac{\rho}{\varepsilon\varepsilon_0} \tag{3.1}$$

Let us turn from the potential $V(x)$ to the potential energy of electrons $\varphi(x)$, $(\varphi(x) = -qV(x))$. Since $\rho = qN_\mathrm{d}$, we shall have:

$$\frac{d^2\varphi}{dx^2} = \frac{q^2N_\mathrm{d}}{\varepsilon\varepsilon_0} \tag{3.2}$$

where N_d is the concentration of the p-type impurities, ε is the relative permittivity of the substance, ε_0 is the permittivity of free space, and q is the electron charge.

Supposing the space charge distribution being stepped, and considering the shielding distance $L_\mathrm{sh} = \sqrt{(\varepsilon\varepsilon_0 kT)/q^2 N_\mathrm{d}}$ and designating $kT = \theta$ (3.2), we shall present eq. (3.1) more conveniently for the solution:

$$\frac{d^2\varphi(x)}{dx^2} = \frac{\theta}{L_{sh}} \tag{3.3}$$

The boundary conditions for the given equation are $\varphi(x) = \varphi_{b1}$ at $x = 0$ and $\varphi(x) = \varphi_{b2} + V$ at $x = d$, where V is the value of the external voltage applied to the structure. Designating x_m for the point in which the electrical field is absent ($|d\varphi/dx| = 0$) and, given the boundary conditions, by integrating eq. (3.3), we shall obtain

$$\frac{d\varphi(x)}{dx} = \frac{\theta}{L_{sh}}(x - x_m) \tag{3.4}$$

By integrating eq. (3.4), we shall obtain

$$\varphi(x) = \frac{\theta}{2L_{sh}^2}x^2 - \frac{\theta}{L_{sh}^2}x \times x_m + C_2 \tag{3.5}$$

At the boundary conditions $\varphi(0) = \varphi_{b1}$, from eq. (3.5), we shall have

$$\varphi(x) = \frac{\theta}{2L_{sh}^2}x^2 - \frac{\theta}{L_{sh}^2}x \times x_m + \varphi_{b1} \tag{3.6}$$

As, in the presence of the external voltage V, the potential energy φ in the point d is equal to $\varphi(d) = \varphi_{b2} + qV = \frac{\theta}{2L_{sh}^2}d^2 - \frac{\theta}{L_{sh}^2}d \times x_m + \varphi_{b2}$ (from eq. (3.6)), it is possible to find the dependences of x_m and $d - x_m$ on the voltage

$$x_m = \frac{d}{2} - \frac{L_{sh}^2(\Delta\varphi - qV)}{\theta d} \tag{3.7}$$

$$d - x_m = \frac{d}{2} + \frac{L_{sh}^2(\Delta\varphi + qV)}{\theta d} \tag{3.8}$$

where $\Delta\varphi_b = \varphi_{b2} - \varphi_{b1}$.

Thus, in the structures where the high-resistance layer is completely covered with the SCR of the oppositely directed barriers from both sides, the width of each of the barriers within the limits of the change from 0 to d depends linearly on the external voltage. In CdTe-based structures, the high-resistance layer is restricted by the surface Schottky barrier and the low-resistance region with the n-type conductivity (the ratio of the specific resistance of the high-resistance layer and the low-resistance n-region is ~10^8). In this case, eqs. (3.7) and (3.8) are also valid if we consider that $\Delta\varphi_b = \Delta E_F - \varphi_b$, where φ_b is the height of the potential Schottky barrier (Figure 3.11) and ΔE_F is the difference in energy between the Fermi level and the bottom of the conduction band in the low-resistance n-region.

From eq. (5.7), it is also possible to obtain the expression for V:

$$V = \frac{qN_d d(d - 2x_m)}{2\varepsilon\varepsilon_0} - \frac{\Delta\varphi}{q} \tag{3.9}$$

The limiting values of the voltage, where $x_m = d$ and $x_m = 0$, can be obtained from eq. (3.9). For silicon samples $V_d = -1.23\,V$ and $V_0 = 1.03\,V$ and for CdTe-based samples $V_d = -0.56\,V$ and $V_0 = 0.16\,V$. Based on the dependence of the point of change of the

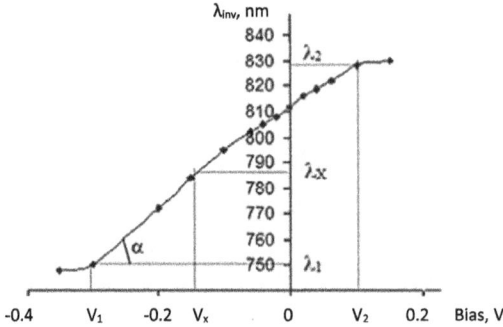

Figure 3.12: Dependence of the point of change of the spectral photocurrent sign λ_{inv} on the bias voltage.

spectral photocurrent sign on the voltage (Figure 3.12), the unknown wavelength of the absorbed beam can be determined by the tangent of the angle of inclination of that linear section ($tg\alpha$):

$$\lambda_x = \frac{V_x(\lambda_2 - \lambda_1) + \lambda_1 V_2 - \lambda_2 V_1}{V_2 - V_1} \tag{3.10}$$

where V_x is the measured bias voltage at which the photocurrent is zero, V_2 and V_1 are the threshold voltage values of the line site in Figure 3.12, λ_1 and λ_2 are the wavelengths corresponding to voltages V_1 and V_2.

The linear plot of the voltage λ_{inv} dependence of the experimental samples of CdTe lies in the range of $\lambda_{inv} \sim 0.75 - 0.84\,\mu m$ (Figure 3.12). The sensitivity of the switching point to the voltage at the specified location is $\sim$6 mV/nm. With measurement accuracy of 1 mV, the spectral attenuation of the test samples is 0.17 nm. In the meantime, for instance, in the best device of Edmund Industry Optic GmbH, it is 0.2 nm. It should be noted that the silicon samples also have a plot of linear dependence of the spectral photocurrent change point on the voltage (Figure 3.6), and with a wavemeter capability. Thus, the investigated structures controlled by external voltage have additional functions, such as wave-measuring capabilities, based on which a wave-measuring device can be created.

3.6 Dark current-voltage characteristics of a structure with a high-resistance thin layer between oppositely directed Schottky barriers

When estimating the current density through the structure, it is necessary to consider both the diffusion and the drift components of the current:

$$j = qn\mu_n E + qD_n \frac{dn}{dx} \tag{3.11}$$

Using Einstein's relation $D = \mu kT/q$, and also taking into account that $E = -(dV/dx)$, let us express the voltage V of the field in terms of the potential φ. For this purpose, let us multiply the two parts of eq. (3.10) by $e^{\varphi(x)/kT}$. As a result, the equation will take the form

$$je^{\frac{\varphi(x)}{kT}} = n\mu_n e^{\frac{\varphi(x)}{kT}} \frac{d\varphi(x)}{dx} + \mu_n e^{\frac{\varphi(x)}{kT}} \frac{dn}{dx} = \mu_n kT \frac{d}{dx}\left[ne^{\frac{\varphi(x)}{kT}}\right] \tag{3.12}$$

Integrating both parts of the equation on x from 0 to d, we shall have

$$j \int e^{\frac{\varphi(x)}{kT}} dx = \mu_n kT \int_0^d \frac{d}{dx}\left[ne^{\frac{\varphi(x)}{kT}}\right] dx = \mu_n kT\left[n(d)e^{\frac{\varphi(d)}{kT}} - n(0)e^{\frac{\varphi(0)}{kT}}\right] \tag{3.13}$$

Hence,

$$j = \frac{\mu_n kT\left[n(d)e^{\frac{\varphi(d)}{kT}} - n(0)e^{\frac{\varphi(0)}{kT}}\right]}{\int_0^d e^{\frac{\varphi(x)}{kT}} dx} \tag{3.14}$$

The boundary conditions for eq. (3.13) are $\varphi(d) = \varphi_{b2} + qV$; $\varphi(0) = \varphi_{b1}$. As $\varphi_{b2} - qV \gg kT$; $\varphi_{b1} - qV \gg kT$, the concentration of the electrons on the opposite surfaces of the structure at the contacts will be equal according to (S. Kh. Khudaverdyan et al., 2005)

$$n(d) = N_c e^{-\frac{\varphi_{b2}}{kT}} \text{ and } n(0) = N_c e^{-\frac{\varphi_{b1}}{kT}} \tag{3.15}$$

In view of these expressions, we shall transform the numerator of expression (3.14):

$$\mu_n kT\left[n(d)e^{\frac{\varphi(d)}{kT}} - n(0)e^{\frac{\varphi(0)}{kT}}\right] = \mu_n kT\left[N_c e^{-\frac{\varphi_{b2}}{kT}} e^{\frac{\varphi_{b2}+qV}{kT}} - N_c e^{-\frac{\varphi_{b1}}{kT}} e^{\frac{\varphi_{b1}}{kT}}\right] \tag{3.16}$$

Equation (3.14) will take the form

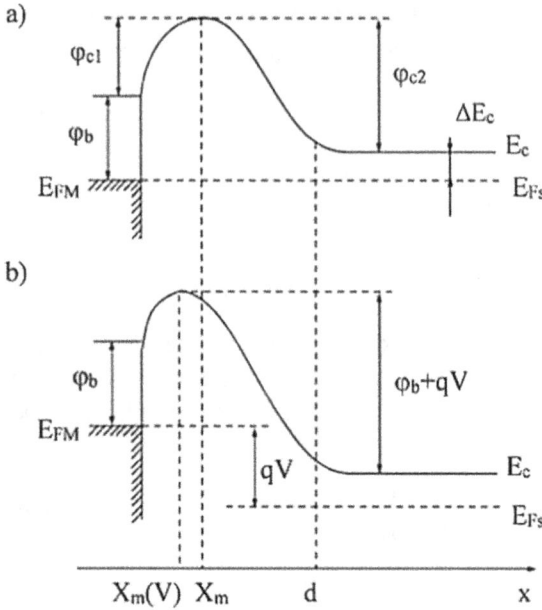

Figure 3.13: Energy band diagram of structures with the surface Schottky barrier and the p-n junction (a) in the absence of the external voltage and (b) in the presence of the external voltage.

$$j = \frac{\mu_n kTN_c \left[e^{\frac{qV}{kT}} - 1 \right]}{\int_0^d e^{\frac{\varphi(x)}{kT}} dx} \tag{3.17}$$

To calculate the denominator in eq. (3.17), we shall divide both parts of eq. (3.6) by KT ($\theta = KT$) and obtain

$$\frac{\varphi(x)}{kT} = \frac{1}{2}\left(\frac{x - x_m}{L_{sh}}\right)^2 - \frac{1}{2}\left(\frac{x_m}{L_{sh}}\right)^2 + \frac{\varphi_{b1}}{kT} \tag{3.18}$$

Substituting the expression $\varphi(x)/KT$ in the denominator of eq. (3.17), we shall obtain

$$\int_0^d e^{\frac{\varphi(x)}{kT}} dx = L_{sh}\sqrt{2}e^{-\frac{1}{2}\left(\frac{x_m}{L_{sh}}\right)^2} e^{\frac{\varphi_{b1}}{kT}} \left(\int_0^{\frac{x_m}{L_{sh}\sqrt{2}}} e^{t^2} dt + \int_0^{\frac{d-x_m}{L_{sh}\sqrt{2}}} e^{t^2} dt \right) \tag{3.19}$$

where $\frac{x - x_m}{L_{sh}\sqrt{2}} = t$.

As a result, we shall obtain the expression for dark CVCs of structures with a high-resistance thin layer:

$$j = \cfrac{\mu_n kТN_cL_{sh}^{-1}\sqrt{2}e^{\frac{1}{2}\left(\frac{x_m}{L_{sh}}\right)^2} e^{-\frac{\varphi_{b1}}{kT}}\left(e^{\frac{qV}{kT}}-1\right)}{\displaystyle\int_0^{\frac{x_m}{L_{sh}\sqrt{2}}} e^{t^2}\,dt + \int_0^{\frac{d-x_m}{L_{sh}\sqrt{2}}} e^{t^2}\,dt} \qquad (3.20)$$

where x_m is set by eq. (3.7).

3.7 Light current-voltage characteristics of two-barrier diode structures with a high-resistance thin layer between oppositely directed Schottky barriers

Generally, the photocurrent flowing through the p-n junction or the Schottky barrier consists of diffusion and drift components. The drift photocurrent is determined by the generation of the carriers in the depleted high-resistance region, and the diffusion photocurrent is determined by the generation of the carriers in the region located at the diffusion length of the minority carriers from the junction. The silicon-based structures have a region depleted by two oppositely directed barriers (Figure 3.10), where the generated photocarriers are separated by the junction fields and create drift photocurrent. The diffusion photocurrent is absent in these structures since there is no region where diffusion processes can occur.

In CdTe-based samples, at longwave radiation, a part of the radiation is absorbed outside the space charge field (outside the narrow high-resistance layer) in the n-region (Figure 3.13), where the minority carriers, photogenerated at the diffusion length, get separated by the junction field via diffusing toward the p-n junction and create the diffusion photocurrent.

Let us deduce an expression for the photocurrent through the silicon structure. The speed of the electron-hole pair generation changes with the change of the coordinate according to the exponential law (S. Khudaverdyan et al., 2005; Sze et al., 2021):

$$G(x) = J_0 a e^{-ax} \qquad (3.21)$$

where J_0 is the flow of the incident photons per unit area in a unit of time equal to $W(1-R)/S \times h\vartheta$. Here R is the reflection coefficient, a is the absorption coefficient, W is the radiation power, and S is the photosensitive surface. The electrons $G(x)$ generated in the point x in a unit time make a contribution equal to the drift photocurrent passing through the field. We shall present the photocurrent density through the first junction as

$$J_{Ph1} = q \int_0^{x_m} G(x)dx = qJ_0\left(1 - e^{-\alpha x_m}\right) \tag{3.22}$$

and the photocurrent density through the second junction as

$$J_{Ph2} = -q \int_{x_m}^{d} G(x)dx = qJ_0\left(e^{-\alpha x_m} - e^{-\alpha d}\right) \tag{3.23}$$

The resulting photocurrent density through the investigated structures can be presented as

$$J_{Ph} = J_{Ph2} + J_{Ph1} = qJ_0\left(2e^{-\alpha x_m} - e^{-\alpha d} - 1\right) \tag{3.24}$$

In view of the expressions for the photocurrent density (3.24) and the dark current (3.20), we shall obtain the general expression of the current density through the structure

$$J_c = \frac{\mu_n kTN_c L_{sh}^{-1}\sqrt{2}e^{\frac{1}{2}\left(\frac{x_m}{L_{sh}}\right)^2} e^{-\frac{\Phi_{b1}}{kT}}\left(e^{\frac{qV}{kT}} - 1\right)}{\int_0^{\frac{x_m}{L_{sh}\sqrt{2}}} e^{t^2}dt + \int_0^{\frac{d-x_m}{L_{sh}\sqrt{2}}} e^{t^2}dt} + qJ_0\left(2e^{-\alpha x_m} - e^{-\alpha d} - 1\right) \tag{3.25}$$

Figure 3.14 presents the CVCs (curve a) for silicon-based structures in the darkness, deduced by eq. (3.20). As the figure shows, there are three different sites on CVCs. To explain the shape of the curve, it is necessary to consider that the current through the structure is basically caused by the largest/highest potential barrier. In the absence of the bias voltage, the rear barrier is higher by 0.1 eV than the surface barrier. In the presence of the external voltage ("+" on the contact of the surface barrier), the surface barrier is biased in the direct, and the rear barrier in the opposite direction is reverse-biased. In this case, the saturation of the current (site 1, Figure 3.14) is caused by prevailing processes of the current passing through the reverse-biased rear barrier. With the change of the polarity of the external voltage, the surface barrier is reverse-biased, and the rear barrier is in the direct direction, and prevailing in the beginning is the passage of the current through the directly forward-biased barrier, which causes the initial growth of the current (site 2, Figure 3.14). With the further increase of the bias voltage, the height of the reverse-biased surface barrier becomes more than that of the forward-biased rear barrier, and the dependence passes to the slower growth of the current (site 3, Figure 3.14).

The CVCs at illumination, built by eq. (3.25), were similar to dark ones. At a certain bias voltage (the rear barrier is reverse biased in the opposite direction, $V > 0$), the dark and at illumination, CVCs are crossed. And the less the wavelength of the ab-

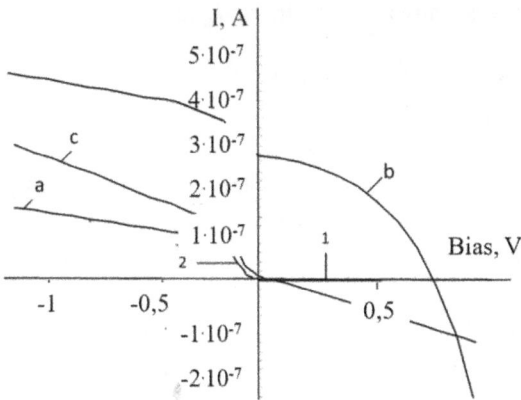

Figure 3.14: The current-voltage characteristics of the silicon-based structures in the dark (curve a) and at illumination with different wavelengths λ: (curve b) 0.4 μm; (curve c) 0.4 μm. Incident radiation power is 1 μW.

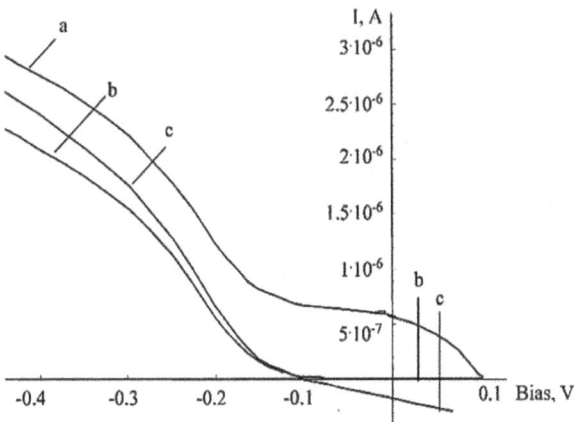

Figure 3.15: The current-voltage characteristics of cadmium telluride-based structures in the dark (curve b) and at illumination with different wavelengths λ: (curve a) 0.6 μm; (curve c) 0.84 μm. The incident radiation power is 1 μW.

sorbed radiation (Figure 3.14, $\lambda = 0.4$ μm at curve b and $\lambda = 0.56$ μm at curve c), the larger the voltage at which the crossing occurs. Such behavior can be explained by the fact that the shorter the wavelength of the radiation, the less the depth of its absorption, and for the photocurrent to be generated in the area of the rear junction, greater bias voltage is necessary, which increases the SCR width of the rear barrier and reduces the SCR width of the surface barrier.

The CVCs for CdTe-based structures in the dark and at illumination (Figure 3.15) are built by eqs. (3.20) and (3.25). The main differences between the CVCs of these structures and the silicon-based structures are the wider range of voltages of the exponential growth of currents and the less expressed transition to saturation at the reverse bias of the surface barrier.

The first difference is due to the bigger difference in the barrier heights (0.2 eV) in CdTe-based structures as compared to that in silicon-based structures (0.1 eV). The second difference is due to the smaller height of the surface potential barrier in the CdTe-based structures (0.4 eV) as compared to that in silicon-based structures (0.5 eV). In addition, the presence of the diffusion current through the p-n junction in CdTe-based structures is ignored. The second difference is due to the small height of the surface potential barrier (0.4 eV) in structures based on CdTe, in contrast to the structures based on silicon (0.5 eV). In addition, in structures based on CdTe, the presence of diffusion current through the transition is not considered.

3.8 Summary

Three features were observed in photodetectors made on the basis of CdTe and Si. Firstly, with increasing voltage, the range of wavelengths of the positive spectral photocurrent increases significantly. Secondly, the point of change of sign of the photocurrent shifts toward longwaves (in the case of a voltage with the opposite shift – toward shortwaves). Thirdly, the point of change of sign of the spectral photocurrent does not depend on the beam power. In both cases, the reverse dark current induced by the large potential barrier is much smaller than the reverse dark current induced by the small potential barrier. These phenomena are explained by the mutual compensation of photocurrents of opposite potential barriers of the structure from external influences.

Part 2: **Methods of spectral-selective sensitivity of optical radiation**

Chapter 4
Spectral-selective sensitivity of optical radiation

4.1 Introduction

Spectral-selective sensitivity in photodetectors refers to the ability of a photodetector to selectively detect light within a specific range of wavelengths or colors. This feature is crucial in various applications where precise detection of certain wavelengths of light is required, such as in spectroscopy, color sensing, remote sensing, and optical communications. There are several methods to achieve spectral-selective sensitivity in photodetectors. (a) Material selection: Different semiconductor materials have varying bandgap energies, which determine the wavelengths of light they can absorb. By choosing appropriate materials, photodetectors can be tailored to detect specific wavelengths. For instance, silicon photodetectors are commonly used for visible light detection, while materials like indium gallium arsenide (InGaAs) are used for near-infrared (IR) detection. (b) Bandgap engineering: By changing the composition and structure of semiconductor materials, the bandgap energy can be tuned to target specific wavelengths. This can be achieved through techniques such as alloying, doping, or quantum well structures. (c) Optical filters: Photodetectors can be equipped with optical filters that selectively transmit desired wavelengths while blocking others. These filters can be integrated directly into the photodetector or placed externally in the optical path. (d) Spectral response calibration: Calibration techniques can be employed to precisely characterize the spectral sensitivity of photodetectors. This ensures accurate detection of specific wavelengths and enables calibration for different applications. (e) Multispectral and hyperspectral imaging: In advanced applications, arrays of photodetectors with different spectral sensitivities are used to capture images across multiple wavelengths simultaneously. This enables the creation of multispectral or hyperspectral images for various scientific and industrial purposes.

Spectral-selective sensitivity in photodetectors plays a vital role in various fields, including astronomy, environmental monitoring, medical diagnostics, industrial quality control, and telecommunications. Advancements in material science and fabrication techniques continue to improve the performance and versatility of photodetectors, enabling more precise and selective detection of light across different wavelengths.

4.2 Detectors with selective spectral sensitivity

Different approaches can be applied to make photodetector systems with selective sensitivity depending on the wavelength: using the Fabry-Perot interferometer (Jerman & Clift, 1991), the diffraction grating (Kong et al., 2001), the photodetector with an optimized resonance layer (Kishino et al., 1991), the photodetector without the light

https://doi.org/10.1515/9783111428024-005

filter (Adachi, 1985), the passive optical unit (Aravanis et al., 2007), and so on. The semiconductor photodetector array is used in them for the registration.

The Fabry-Perot interferometer is an optical element that consists of two parallel partially reflecting mirrors. They are separated by a slit that can be used as a wave filter (Jerman & Clift, 1991). When the slit is equal to the integer multiple of the half-wave of the incident light, a resonance transmission peak occurs. The output signal of the Fabry-Perot filter is a function of the angle of the incident beam. It is a drawback. The characteristics of the filter according to the wavelengths change when the incident beam deviates from the normal. Another way to select waves is to use a diffraction grating. It is used to disperse or scatter the incident light for the spatial distribution of waves (Kong et al., 2001). In this case, too, the diffraction image changes as a function depending on the angle of the incident beam.

The photodetectors with an optimized resonance layer are photonic devices that ensure the selective spectral sensitivity of the radiation. The selectivity of the device increases due to the installation of the active layer of the device in the Fabry-Perot microlayer. Here, too, the resonator's quantum efficiency depends on the incident light's angle. When the angle of the incidence deviates from the normal, the value of the maximum quantum efficiency changes since the change in the optical path length results in a change in the resonance wavelength (Kishino et al., 1991).

There were also attempts to create photoreceivers without filters. They could select narrow layers of the optical spectrum. The most popular ones are cascade photodetectors with different base thicknesses (Aravanis et al., 2007; Kalkhoran & Namayar, 1997; Nader & Fereydoon, 2004) and multicolor photodetectors (Gergel et al., 2006; Vanyushin et al., 2005). In the former case, the cascade photodetectors with different base thicknesses are used to control the wavelength (Figure 4.1).

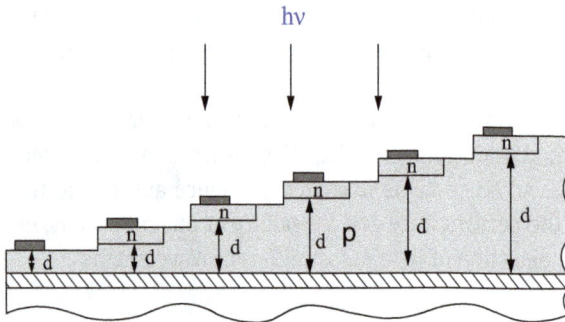

Figure 4.1: Cascade photoreceivers detecting the wavelengths.

The difference in the photocurrents of the photoreceivers of the cascade series is conditioned by the difference in $\Delta\lambda$ of the wave group absorbed in the photoreceiver with the wide base and $\Delta\lambda$ of the wave group partially absorbed in the photoreceiver with the narrow base. The minimum $\Delta\lambda$ is restricted by the threshold sensitivities of photo-receivers. The closer the base widths, the smaller the $\Delta\lambda$. Thus, the spectral selectivity is close to the single channel. Here, there emerge some difficulties in the production technology.

In the latter case, high-resistance layers equal to the colors embedded at different depths from the surface are used (Figure 4.2). It creates technological difficulties caused not only by the number of layers but also by the fact that their mutual transition must be abrupt. Otherwise, the spectrum of the recorded beam will be blurred. The selectivity of the waves is determined by the registration of the wave groups absorbed in the high-resistance layers. The photodiodes made in the semiconductor $(Al_xGa_{(1-x)}As)$ have also been implemented for the selective sensitivity of the wave (Campbell et al., 2000; Seymour et al., 2014a). The main advantage of using this material is that $(Al_xGa_{(1-x)}As)$ is a direct band semiconductor. It has a greater absorption coefficient than silicon in the near-IR region; hence, it has high quantum efficiency, and the output current transformation via radiation absorption.

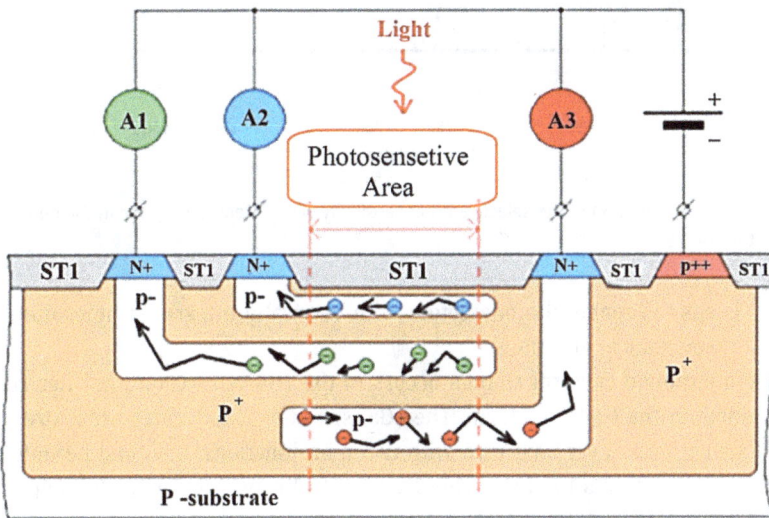

Figure 4.2: Tricolor photoreceiver structure.

In the multilayer $(Al_xGa_{(1-x)}As)$ photodetector structure, the wavelength selectivity is achieved by changing the aluminum and gallium portions in the GaAs/AlGaAs semiconductor system, which leads to the change in the prohibited zone of the AlGaAs, and, hence, to the change in the spectral peak.

In the heterostructure, a section of the large prohibited zone is intended for filtering the short-waves, and the active layer of the photodiode is designed to reduce the sensitivity of certain waves. Figure 4.3 shows the spectral distribution of the selective photosensitivity with a stepwise change in the density of aluminum in steps of 4% (Seymour et al., 2014a).

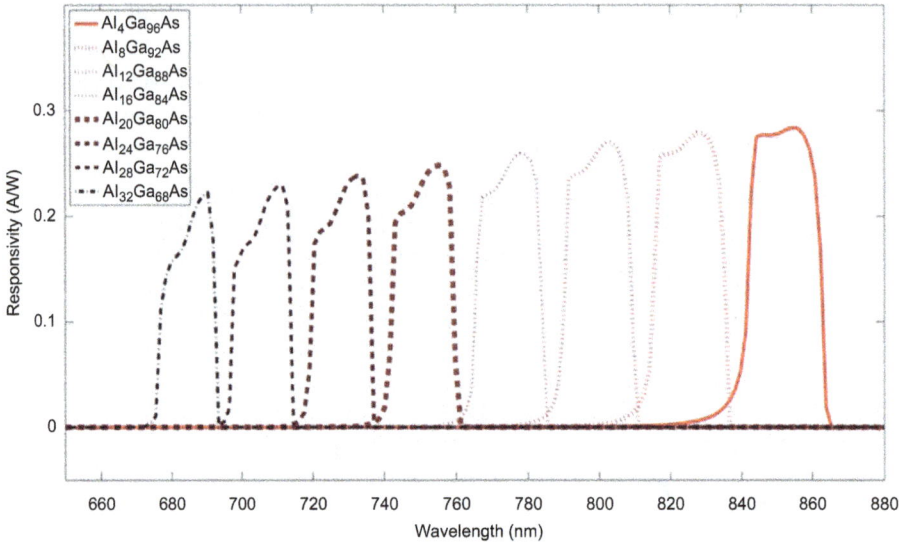

Figure 4.3: The spectral distribution of the selective photosensitivity with a stepwise change in the density of aluminum in steps of 4%.

In addition, it is easy to make photodetectors with vertically integrated heterojunctions (Figure 4.4; Seymour et al., 2014a).

The above-mentioned spectral change occurs in the $(GaAs/Al_xGa_{(1-x)}As)$ heterostructure obtained on the GaAs platform. The $(GaAs/Al_xGa_{(1-x)}As)$ heterostructure is an epitaxially grown multilayer structure with two p-i-n junctions. It should be noted that there were also attempts to obtain selective spectral sensitivity with the help of other types of semiconductor photodetectors, particularly in the shortwave region, using wide-gap semiconductors as the initial material. These are mainly the photodiodes with the Schottky barrier. Figure 4.6 presents the spectral characteristics of the Cr-4H-SiC structure. It covers the wavelength range of 24–0.3 μm (Blank & Gol'dberg, 2003; Campbell et al., 2000). Figure 4.7 presents the spectral characteristics of another structure (Au-ZnO; Campbell et al., 2000).

In these structures, the longwave drop is conditioned by the width of the prohibited zone of the semiconductor, and the shortwave drop is conditioned by high-speed surface recombinations. Together these factors create a rather narrow range of spectral photosensitivity, and the spectral selectivity relates to that part of the spectrum and does not cover wide ranges.

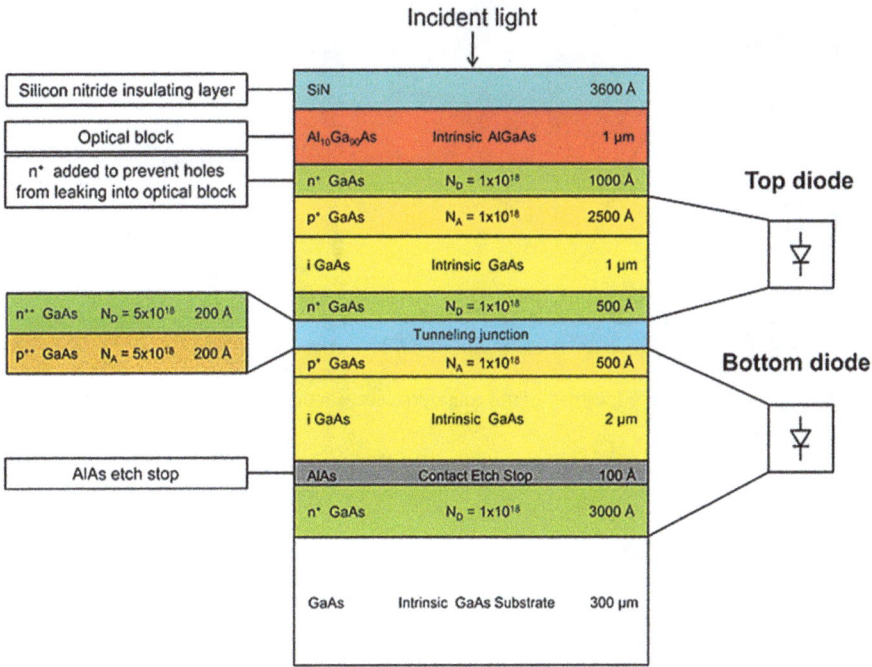

Figure 4.4: The diagram of the cross section of the multicolor photodetector.

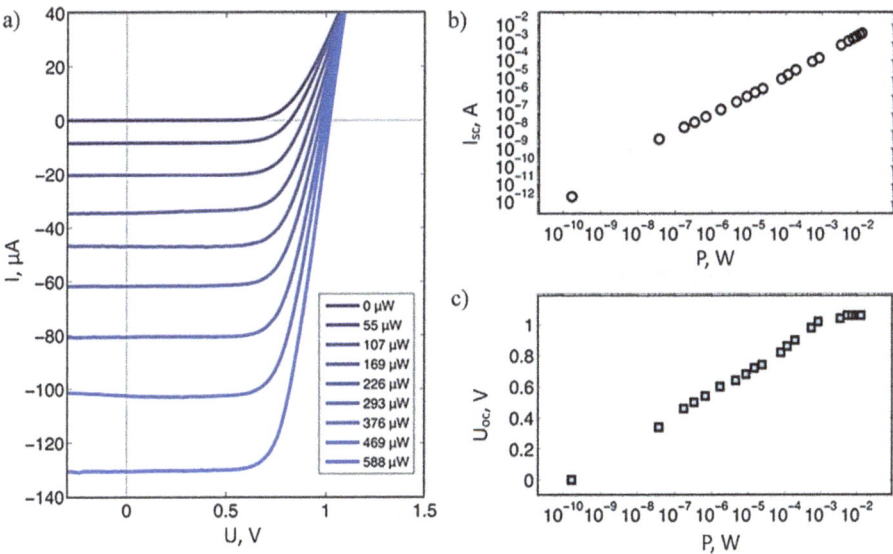

Figure 4.5: (A) The current-voltage characteristic of the structure at different radiation powers of the laser with the wavelength of 856 nm. (B) The change in the short-circuit current depends on the radiation power. (C) The dependence of the open-circuit voltage on the radiation power.

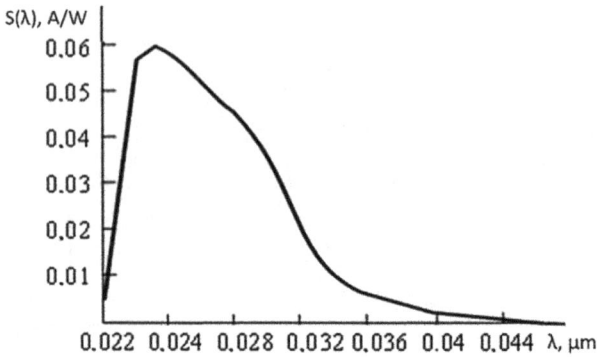

Figure 4.6: The photosensitivity spectrum of the photoreceivers with the Cr-4H-SiC structure with the Schottky barrier at $T = 300$ K.

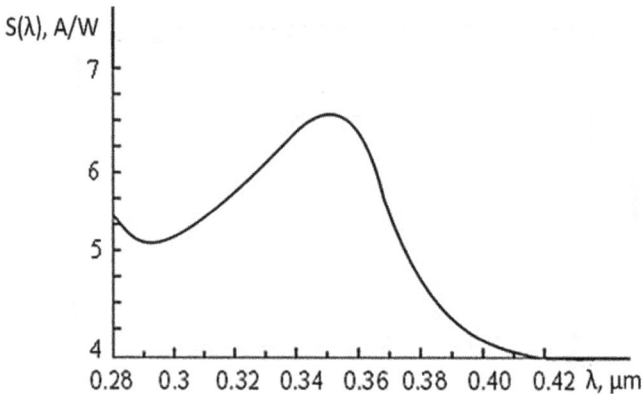

Figure 4.7: The photosensitivity spectrum of the photoreceivers with the Au-ZnO Schottky barrier at $T = 300$ K.

4.3 Spectrophotometric properties of photodetectors with two-barrier structures: advantages and disadvantages

The disharmony between man and nature on the Earth is due to natural and anthropogenic factors. The natural factors, caused by cosmic interactions, affect the Earth slowly, while the anthropogenic ones aggravate the environmental crisis in man-nature relations from day to day. One of the ways to avert the crisis is to create up-to-date monitoring techniques and to implement them to reveal the negative effects of industry, agriculture, and transport, as well as to detect contamination of the sweet water with petroleum, phenols, nitrogen, pesticides, and heavy metals. It is done by monitoring the corresponding environment. The new generation sensors, used for the local detection of contaminants, play a significant role in monitoring systems.

Arustamyan et al. (2010) and S. K. Khudaverdyan et al. (2011, 2016) studied a new model of the two-channel, parallelly distributed photodetector (Figure 4.8) used for the remote identification and quantitative analysis of hazardous substances in the optically transparent environment using optical information. Arustamyan et al. (2010), S. Khudaverdyan et al. (2009, 2012), S. Khudaverdyan, Avetsiyan, et al. (2013), S. Khudaverdyan, Khachatryan, et al. (2013a), S. Khudaverdyan et al. (2017), S. K. Khudaverdyan et al. (2011), S. Khudaverdyan and Vaseashta (2013), and Vaseashta et al. (2013) studied the photospectrometric properties of the photodetector structures with vertically distributed oppositely directed potential barriers (Figure 4.9). To carry out the sequential registration of waves in the structure, the matching of the penetration of longwaves and the depths of the active environment for the possibility of the spectral analysis of the radiation have been provided.

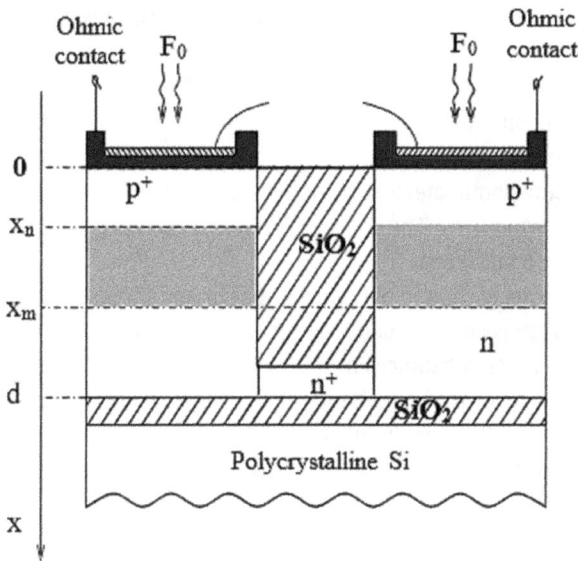

Figure 4.8: Two-channel photodetector structure.

The test sample was prepared to investigate the electrophysical and photoelectric properties of the n^+-p-n^+ structure with an epitaxially grown base, and the spectral-selective sensitivity has been investigated. A monitoring system equipped with a photospectrometric unit has also been developed. It ensures the processing of the information obtained via modern telecommunications and the digital modeling of the processes.

The mathematical expressions connecting the structural parameters of the photodetector, the photocurrent, the spectral composition, and the intensity of the beam have been obtained. They serve as the basis for the development of a preliminary ver-

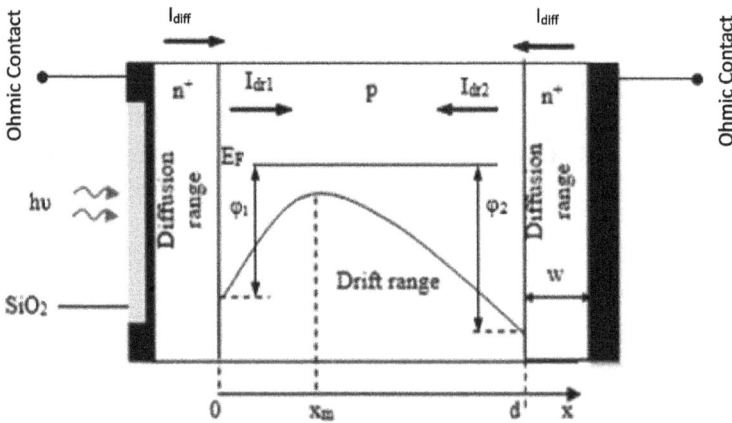

Figure 4.9: n^+-p-n^+ structure displaying the directions of the currents (figure from our earlier work is printed with permission).

sion of an algorithm for the identification and quantitative analysis of the mixtures in the optically transparent environment.

The investigated semiconductor photodetectors have been studied in terms of the detection of photospectrometric properties. The following issues regarding the reliability of the results have not been considered:

– The coordinated choice of the design and technological parameters of the photodetector for obtaining accurate results, depending on the requirements of the identification of hazardous or other substances in this or that applied problem.
– The accuracy of the spectrum reconstruction depends on the absorption coefficient a, and on the inaccuracies of the transition from a to λ.
– To obtain more accurate values of the absorption coefficient, it is necessary to discard the approximate expression and solve the transcendental equation, which should be further included in the algorithmic process.
– The diffusion layer width, potential barrier height of silicon structure, and surface transparency for shortwaves.

4.4 State-of-the-art capabilities of the optical spectral analysis

As it was mentioned above, spectral analysis can solve a wide range of problems: the analysis of very-high-purity materials, the inspection of finished products, the rapid analysis of alloys in metalworking production, the exploration of mines for mineral deposits, the analysis of the composition of the Moon surface and stellar materials, the monitoring of the industrial and household running water, the control of the air pollution of production areas, and the solution of agricultural, medical, physical, and other problems. These problems need monitoring analysis, and the proposed photodetector can be applied in the optical monitoring systems as a photosensor of the primary information which, using inexpensive electronics, is capable of analyzing the information on the optical signal, identifying hazardous and other substances in the investigated medium, determining their quantity, sending the information wirelessly, eliminating potential dangers, thus solving the security problems. In these processes, the crucial parameters of the proposed detector are small size, high accuracy, high speed, and low price. They ensure the use of the detectors for monitoring large areas where a large number of sensors is required, like the monitoring of drinking water or other water resources. It is known that for the detection of heavy metals in water, it is necessary to register the electromagnetic waves of the ultraviolet (UV), visible, and near-IR regions. These regions correspond to the spectral sensitivity range of the proposed photodetector.

The sensor with the spectrophotometric system is quite small and therefore can be applied in field investigations. Some types of toxins emit fluorescent radiation into the UV region: $B1$ alpha toxins in 265 and 362 nm, $M1$ alpha toxins in 265 and 357 nm, and sterigmatocystin in 324 nm (Ball, 2006; Egmond et al., 2004; K. Grigoryan, 2008; Joint FAO/WHO Expert Committee on Food Additives. Meeting (68th : 2007 : Geneva et al., 2007; Lewis et al., 2005). If the silicon detectors have a window in the UV region of the spectrum, it is possible to quantitatively identify those toxins and determine the safety level.

Photospectrometry is the measurement of the intensity of the electromagnetic radiation reflected, absorbed, or passed through the material, depending on the wavelength (Ball, 2006; G.E., 2013; Owen, 2000; Saptari, 2004; Shimadzu Corp., 2017; Stuart, 2008; Trumbo et al., 2013). In practice, the most commonly used spectral range of electromagnetic radiation includes the UV (190–380 nm), the visible (380–780 nm), and the near-IR (780–2,500 nm) regions. Figure 4.10 presents the visible light spectrum (Owen, 2000).

Generally, the following principles are applied in photospectrometry (Owen, 2000):

1. When studying a solid body, the radiation with the previously measured intensity is passed through the test sample. Then, the intensity of the radiation that comes out of the sample is measured. The spectral difference between the input and output intensities gives information on the type and the number of absorbed particles in the solid body.
2. When studying the solutions, first, the spectral distribution of the radiation intensity passing through the pure solvent is measured, and then the spectral distribution of the radiation intensity coming out of the solution containing the mixture. Their difference gives information on the type and the number of the absorbed particles in the solution.

Modern photospectrometers, as was already mentioned, ensure the single-channel radiation input into the sample. It happens due to the dispersion of the multiwave radiation. The dispersion is realized either with the help of a prism or a diffraction grating.

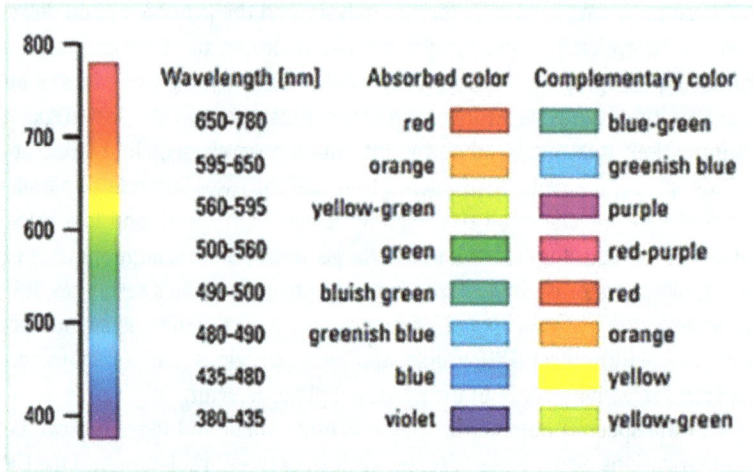

Figure 4.10: Visible light spectrum: absorbable and additional colors.

Light filters are rarely used. They may not transmit shortwaves (except for the filtered wave), but they transmit IR radiation (Blank & Gol'dberg, 2003). Besides, the filtering is not strictly single channel, and the transmission spectrum has a slit $\Delta\lambda$. That is why they are not effective. In the case of a prism (Figure 4.11), the wave-dispersed radiation directed through the narrow slit turns into a monochromatic wave and falls onto the surface of the test sample, and is then recorded by the photodetector (Owen, 2000).

The change in the waves absorbed in the sample is carried out using the high-precision rotation of the prism. It is also possible to place the test sample before the dispersion. In that case, a linear series of photodetectors or an array of photodetectors with charge coupling is placed against the dispersed radiation for the registration of individual waves. The spectral accuracy reaches 1 nm.

The same accuracy is provided during the dispersion with the diffraction grating. In this case, too, the sample is placed against the radiation before the dispersion, and a linear series of photodetectors or an array of photodetectors with the charge coupling is placed against the dispersed radiation (Figure 4.12). It is also possible to place the sample after the dispersion. Although these variants of methods dominate the market today, they contain sophisticated and expensive optical and mechanical devices. As it was already mentioned, monochromators and spectrophotometers are mainly used for optical spectral analysis. The dispersion (the scattering) of the white light in them is realized using prisms (Figure 4.11), and now, more often by diffraction gratings (Figure 4.12; Mogniotte et al., 2018).

The diffraction gratings have equally spaced parallel lines from several hundreds to about 2,000 on the surface of 1 mm^2 (Figure 4.11). Because of the interference, the beam of the white light scatters through the diffraction grating perpendicular to the lines, and the light component with a certain wavelength is reflected only at a specific angle.

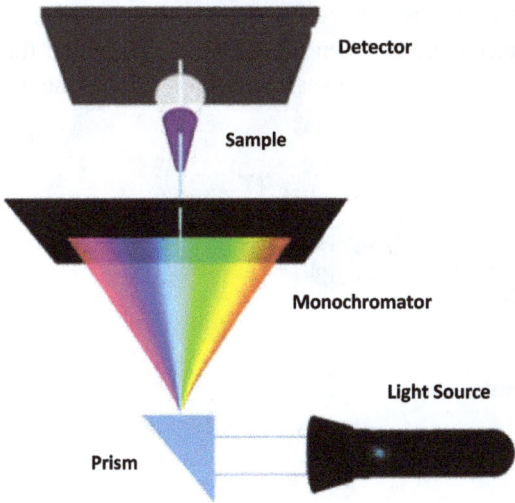

Figure 4.11: The structure of the photospectrometer with the prism.

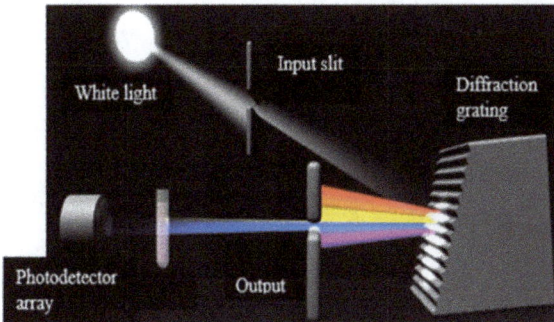

Figure 4.12: Light scattering (dispersion) by the diffraction grating.

As a result, the beam scattered through the grating has a certain deflection where the waves are arranged in a sequence of colors. If an output slit transmitting one color is placed in front, then the wave passing through the sample will transmit the information on its absorption toward the photodetector array (Figure 4.12). In that case, to register the information of all the waves, it is necessary to rotate the diffraction grating with the help of an accurate mechanical system so that, every subsequent moment, a new wave can pass through the slit and, therefore, through the sample. Sometimes, the sample is placed immediately against the scattered beam, and then, an array of the photodetectors is placed at a large distance if the high spectral resolution is required, and at a small distance if the accuracy requirement is not high. In this case, the system rotation is not needed.

In Figures 4.13 and 4.14, a prism is used as a scattering element (Owen, 2000). In Figure 4.13, the sample is placed immediately between the radiation source and the input slit, and in Figure 4.14, the sample is placed between the output slit and the detector after the dispersion of the beam.

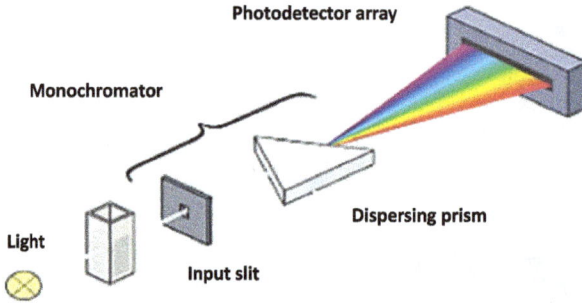

Figure 4.13: The simplified structure of the photospectrometer with the array of photodiodes. The sample is placed between the radiation source and the input slit.

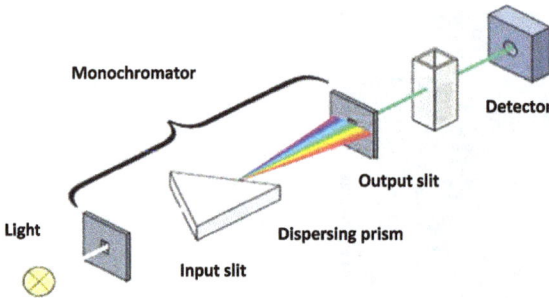

Figure 4.14: The typical diagram of the two-path photospectrometer.

The detector can be mainly a linear series of silicon photodiodes, an array, or an array of charge-coupled devices (CCDs). In the first case, the prism is in the fixed position; in the second case, it makes a high-precision rotation.

4.4.1 Light sources

The basic properties of the light sources used in photospectrometers are (Shimadzu Corp., 2009):
1. The intensity of the wavelengths in the wide range
2. The time stability of the intensity

3. The expiration date
4. The low cost

At present, halogen lamps are often used in the visible and near-IR region of electro-magnetic radiation, and deuterium lamps in the UV region. Xenon and other lamps are used in certain cases. In **halogen lamps** (OSRAM, 2002), the radiation is obtained by the electric heating of the spiral of the tungsten conductor and the delivery of the beam (Figure 4.15).

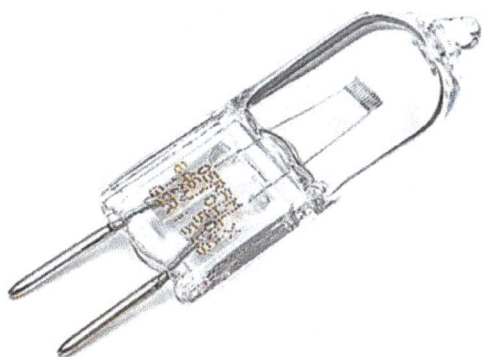

Figure 4.15: An example of a halogen lamp (OSRAM, 2002).

The halogen lamp is filled with inert gas containing a small amount of halogen. The latter contributes to the return of the evaporated tungsten to the filament. It increases the life span by about 2,000 h, which is a high rate for the low prices. Figure 4.16 shows the approximate spectral distribution of the radiation intensity at 3,000 K.

The deuterium lamp (Figure 4.17; Medremkomplekt Company, 2002) is a dis-charge arc light source. It contains deuterium under pressure of several hundred pas-cals (D_2). It is used in various devices (monochromator, spectrophotometer, etc.) as a steady source of UV radiation within the wavelength range of 186–360 nm. In the shortwave range, the windows made of special materials transmitting waves are used. In such photospectrometric devices, the array consisting of photodiodes meas-ures the intensity of each wavelength of the light using a separate detector placed in-side it.

For silicon detectors, the high operation speed is ensured (Owen, 2000), even if multiple measurements of the absorption spectrum for one sample are carried out, and then the output spectrum is obtained by averaging these measurements.

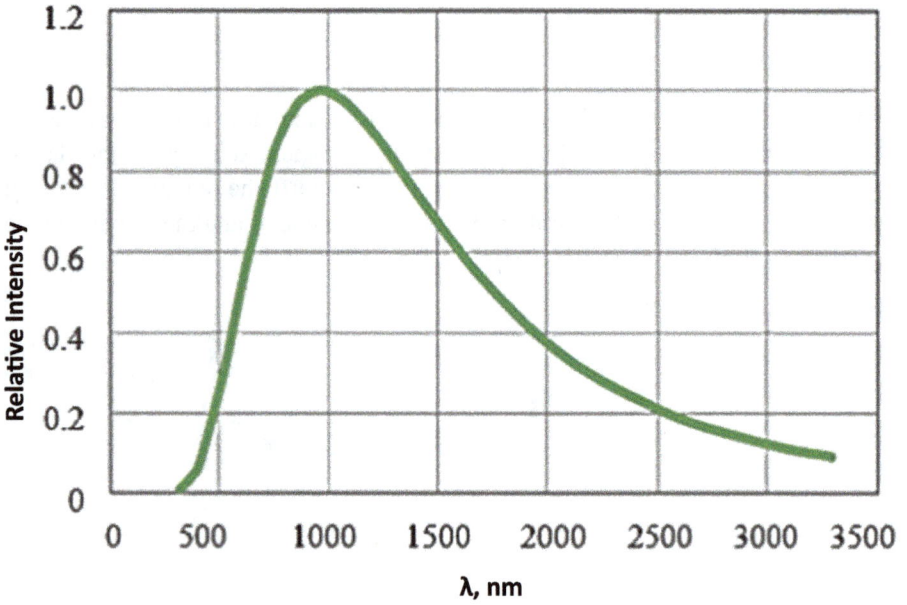

Figure 4.16: The spectral distribution of the radiation intensity of the halogen lamp (3,000 K).

Figure 4.17: An example of a deuterium lamp (Medremkomplekt Company, 2002).

4.4.2 Detectors

The photomultipliers and the semiconductor (e.g., silicon) photodiodes are mainly applied for the registration of UV and visible radiation (Figure 4.18; Owen, 2000). As a result of the radiation absorption, the photocurrent is generated in the silicon (Si) photodiode. If the energy of the photons of the absorbed radiation is greater than the energy of the prohibited zone, then the electrons of the valence band within the entire absorption range (i-layer), obtaining the required energy, pass into the conduction band leaving holes in the valence band.

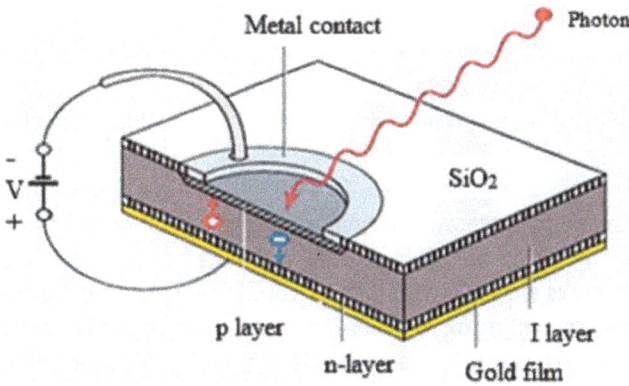

Figure 4.18: The structure of the silicon (Si) photodiode.

The photomultipliers require a high-voltage power supply and have threshold photo-sensitivity, that is, they feel the absorption of a single photon.

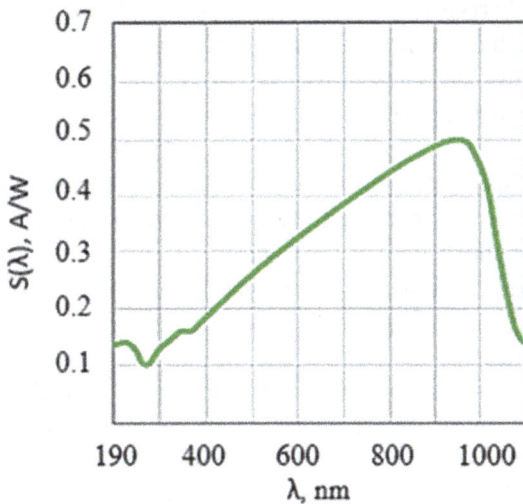

Figure 4.19: The spectral sensitivity of the silicon photodiode.

If the detector is connected to an electrical circuit, the photocurrent is generated (Owen, 2000). The width of the prohibited zone of the silicon (the energy) is approximately 1.12 eV; thus, the upper limit of the spectral sensitivity of the structure is approximately 1,100 nm (Figure 4.19; Shimadzu Corp., 2017).

As compared with the latter ones, the silicon photodiodes have several advantages:
1. Small size, high speed
2. Low price, reliability
3. They do not require a special power supply
4. The structure is technologically compatible with the technological cycle of microcircuit production

The CCDs are the circuits made by integrated technology. Using the voltages applied to the gates sequentially arranged on the surface of the silicon epitaxial layer, they form areas (cells) – pocket arrays enriched with the electrons generated from the absorbed light under the gates. Depending on the voltage magnitude, they are enriched or depleted from the charge carriers that contain the information on the intensity of the absorbed light. The corresponding signal is amplified, digitized, and read at the output. By changing the voltages applied to the gates, the heights of the potential holes are changed, and the hole-to-hole movement of the charges takes place, which eventually leads them to the output of the device (Figure 4.20, step-by-step from a to b and c). The output signal is transformed into voltage and amplified. It gives the information on the intensity, which is digitized and stored (Hosch, 2023).

The arrays of CCDs simultaneously register all the waves incident on the detector and do not require the selection of an individual wave from the light dispersion.

At the same time, the arrays of CCDs have obvious disadvantages:
– Sophisticated production technology, the wide surface occupied on the crystal, and, thus, a high cost.
– The incompatibility with the technological cycle of the production of complementary metal-oxide-semiconductor circuits, and the complicated power supply system when compared with certain photodiodes.
– Complex control system, since the periodical recharging of pockets is required.
– The multistage control over the heights of potential holes is realized using the timing signal and a complex system.
– In the charge reading process, the registration of the data of the charge amplifier and the data taken from the array toward the shift register generates noises and, at high light intensities, due to the abundance of charges, it propagates toward the neighboring pixels and, together with residual charges, distorts the registration results.
– It is often required to reduce large dark currents. Therefore, cooling systems have to be used.
– The information bit is obtained using the sequentially shifted register. If high array resolution is required, the process takes a long time.

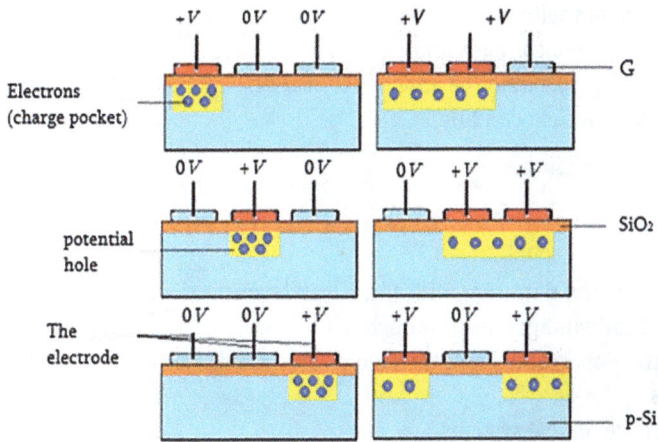

Figure 4.20: The order of the movement of charges in the CDD.

In the near-IR region, InGaAr semiconductor photodiodes have been used in recent years. They are photovoltaic cells with a p-n junction. They have lower energy of the forbidden band than silicon, and, therefore, they absorb longer waves (Figure 4.21) than silicon. They are used as detectors for near-IR regions in spectrometry (Shimadzu Corp., 2017).

In modern photospectrometry, the Fourier transform method is used for the average IR region (3–10 mm). The result of measuring the IR beam transmitting through the sample undergoes Fourier transform, and the information on the spectrum is obtained. It is done a dozen times faster than in the case of the dispersion with the diffraction gratings. This method is widely applied in organic synthesis, polymer science, petrochemical engineering/industry, the pharmaceutical industry, and the spectral analysis of food (Becker & Farsoni, 2014; Ocaya, 2016; Ploykrachang et al., 2011).

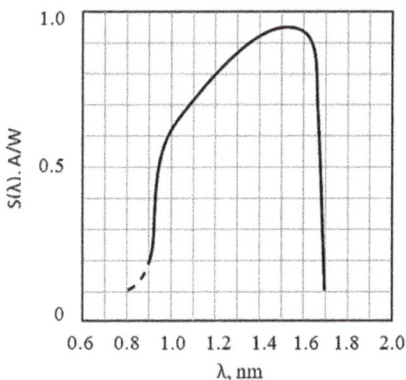

Figure 4.21: Spectral distribution of InGaAs photodiode sensitivity.

Although the IR spectrometers with the Fourier transform have advantages over the other spectrometers, they have a complex and expensive optical system, and they are heavy and large. Therefore, they cannot be applied for real-time monitoring and measuring, and remote analysis in field conditions.

4.5 Summary

The above-stated studies of the detectors with the selective spectral sensitivity show that they have several disadvantages, even though the photodetectors with the selective sensitivity, the diffraction grating, and the Fabry-Perot interferometer are of single channel:
– Complex and expensive optical systems.
– The output signal in them is a function of the angle of the incident beam. When the incident beam deviates from the norm, the wavelength characteristics change.
– It is hard to use them in the field conditions.
– The above-stated disadvantages do not concern the semiconductor photoreceiver structures with the selective spectral sensitivity that can select narrow bands of the optical spectrum. They are not of single channel and do not ensure selectivity within the wide range of the spectrum.

The most popular photodetectors are the cascade photodetectors with different base thicknesses and multicolor photodetectors. The cascade photodetectors with different base thicknesses can be of single channel within the wide range of the spectrum if the voltage change and the necessary structural parameters are ensured.

The spectral selectivity in multicolor photodetectors can be ensured only by a certain slit $\Delta\lambda$, since the absorption bands providing colors have a certain width where a group of waves with approximately similar lengths is absorbed. In this case, it is important for the intensity of the absorbed radiation to increase or remain the same with the increasing wavelength. Otherwise, the shortwave can have a higher intensity, and therefore, a bigger penetration depth than the long one, and can be absorbed into the layer from which the information on the longwaves only has to be obtained. The above-stated structures have several disadvantages:
– It is necessary to create multiple or multilayer photodiodes. Thus, it is material consuming.
– It is necessary to ensure the identification of the incidence angle on the photosensitive surfaces of photodiodes to preserve the spectral composition, power, and registration accuracy of the incident beam.
– The spectral selectivity is ensured by technological and structural interference and cannot be controlled by the external voltage.
– The complexity of the production technology.

- The selectivity in semiconductor photodetectors is limited by the spectral $\Delta\lambda$ layer. Thus, it is not a single channel.

Despite the obvious advantages such as size, weight, price, and operation speed, the investigated semiconductor photodetectors have been studied to reveal the level of the functional properties and require the discussion and the solution of several problems:

- The registration of individual waves from the integral flux of the absorbed radiation largely depends on the constructive and technological parameters of the photodetector, thus highlighting the importance of the mutually agreed choice of the parameters based on the requirements of the identification of harmful and other substances in the applied problem in question.
- The registration accuracy of the detectors providing the spectral-selective sensitivity is based on the cause-and-effect relations, and it is necessary to reveal their role in the electronic processes.
- The spectrum reconstruction error is particularly due to the inaccuracy of the transition to the wavelength λ from the absorption coefficient α. The $\alpha = f(\lambda)$ data of the 10-nm step of the calculation table are used for silicon. To increase the accuracy, it is necessary to use more precise data from the table.
- The absorption coefficients are determined using simplified expressions. To obtain high-accuracy results, it is necessary to solve a transcendental equation.
- To increase the selective spectral sensitivity of the detector, it is necessary to develop a structure in which the junction point of the barriers is in the position corresponding to the absorption depth of the most deeply penetrated wave, and the widths of the barriers are effectively modulated by the external voltage.
- The surface barrier should be transparent to the shortwaves. It will enable Si to register the wavelengths starting from 200 nm. Therefore, a silicon layer of the barrier with the required maximum height and the required width should be used.
- The diffusion layer below the rear barrier should be narrowed. It will bring a decrease in the registration of the deeply absorbed waves in the layer, thus weakening the influence of the diffusion current.
- It is necessary to obtain and investigate a double barrier structure with an n-base and a platinum silicon potential barrier in which the surface barrier will be high, and the high-mobility electrons compared to the p-base holes will ensure high speed.
- From this point, it is necessary to substantiate the creation of the $p^+(PtSi)$-$n(Si)$-$p^+(Si)$ structure with optimized parameters and technological characteristic properties and to investigate the photoelectronic processes occurring in them.
- It is vital to reveal the principles of the change in the widths of the contacting barriers depending on the external voltage and the impurity density in the base, and to show the connection between those principles and the selective photosensitivity of the structure.

- It is necessary to analyze the possibilities of the efficient registration of the individual waves from the integral flux of the radiation and the possibilities of the determination of the lengths and the intensities of the waves.
- It is imperative to substantiate the need for the investigated structure for creating an inexpensive, high-speed, small-size system of optical analysis fit for use in the field conditions; to study the selective spectral sensitivity by obtaining the spectra of the well-known light sources; and to offer the solutions to the problems encountered in the process.
- It is necessary to analyze the possibilities of the application of the investigated photodetectors in the monitoring and multipurpose spectrophotometric systems to obtain the information from the investigated range.
- To take the following steps, it is necessary to realize the algorithm.
- It is required to carry out the filtering of the digital data of the current-voltage characteristics using the Excel software application, and to keep the ones that conform to the principles of the monotonic increase and decrease of the current, depending on the polarity of the voltage.
- It is necessary to obtain and solve the transcendental equation for the filtered values of CVCs to determine α. Insert the solution into the Excel application to obtain an algorithm for the spectral dependence of the radiation intensity.
- It is essential to investigate the algorithm for the $F_0(\lambda)$ characteristics, to detect the vulnerabilities, and to find solutions to eliminate the problems.

Part 3: **Structural and technological aspects**

Chapter 5
Structural and technological aspects
of photodetectors

5.1 Introduction

Remote photospectrometric sensors have been of particular interest in the last decade. That is due to the need for the spectral analysis of the optical signal in nonstandard situations. The sensors provide necessary information on the composition of the investigated environment and solve important identification issues from a security perspective. Here, it is important to obtain high-speed real-time data without direct contact with the measured object (Kishino et al., 1991; Kong et al., 2001; Stuart, 2008). However, the studies of natural objects utilizing the available resources are not complete from both quantitative and qualitative perspectives. Even in the advanced research methods of optical spectral analysis, there occurs a large scatter of the spectral data of similar objects, which often eliminates the possibility of comparing the measurement results. The processing of the spectral data is derived from its quantitative analysis, and the reliability of which depends directly on the parameters of the current equipment. A large number of sensors are required for monitoring the large areas. Therefore, it is urgent to develop small-size, inexpensive sensors with high-spectral sensitivity fit for remote identification in the field conditions, and to work out algorithms for the accurate registration of the information using these sensors.

5.2 Structural features of the photodetectors derived for photospectrometry

In the current methods of spectral analysis, the spectral distribution of electromagnetic radiation is obtained using light filters, prisms, diffraction gratings, as well as high-precision mechanical devices. The spectrophotometric systems based on them lack universality and require new, additional devices and external software support to perform each new function, which makes them quite expensive and unfit for use in field conditions. One of the most attractive ways to solve the problem is the development of spectrophotometric semiconductor photodetectors aimed at carrying out optical spectral analysis without the use of high-precision mechanical devices, light filters, prisms, and diffraction gratings. The development is still at the research level since the production engineering is rather complicated and special working conditions are required.

The need for the development of photodetectors is based on the market demand (Maida, 2016; Technavio, 2017). Particularly, there is a significant need to develop up-to-date multipurpose monitoring systems to obtain relevant information on the com-

https://doi.org/10.1515/9783111428024-006

position of the investigated environment and to address the identification issues that are important from the perspective of security (Becker & Farsoni, 2014; Brčeski & Vaseashta, 2021; Monea et al., 2017; Ocaya, 2016; Vaseashta, 2021; Vaseashta et al., 2022; Vaseashta & Maftei, 2021).

Thus, the creation of an analytical system based on the semiconductor detector, with spectral-selective sensitivity and high spectral resolution, and its implementation in various spheres will be a great achievement, since it will be cheap, fast-acting, user-friendly, and fit for use in the field conditions. The solution to most of the problems involves the UV and visible ranges of the spectrum of electromagnetic radiation. It is known that in linearly arranged semiconductor double-barrier structures, at longitudinal absorption of the radiation, the redistribution of the absorption of the narrow wave ranges of the integral flux of the radiation between the barriers caused by the external bias voltage can occur (Arustamyan et al., 2010; S. Khudaverdyan, Avetsiyan, et al., 2013; S. Khudaverdyan et al., 2009, 2012, 2017; S. Khudaverdyan, Khachatryan, et al., 2013b; S. K. Khudaverdyan et al., 2011, 2016; Lewis et al., 2005; Vaseashta et al., 2013). It requires the appropriate structural solutions:

1. To provide a transparent surface barrier and minimum surface recombination conditioned by its electric field, and the photosensitivity to the UV radiation for the waves that are longer than 200 nm.
2. To ensure a linear change in the width of the barriers, one at the expense of the other. To provide the width of the absorbing medium, that is, the base width, required for the solution of the target problem.
3. To provide a planar structure consistent with the integrated technology by choosing silicon as a starting material.

Taking into account the above-mentioned problems, a planar version of the photodetector structure is proposed. The simplified cross section of the structure is shown in Figure 5.1. The high-density n-Si base with thickness d contains a metal-n-Si silicide barrier on the surface and a joint area of the p-n potential barrier on the rear side. The high-alloy p^+ layers providing p^+-Si diffusion and a boundary ohmic planar contact are adjacent to the rear side. The structure also contains an n^+ contact output layer and a p^+ protective layer, a SiO_2 layer passivating the surface, a PtSi silicide contact, and an ohmic contact providing the sequential distribution of titanium-tungsten (TiW) and Al membranes. The photosensitive surface is covered with an antireflection SiO_2 layer transparent to UV radiation. This is the way to ensure the vertical arrangement of the potential barriers where the longitudinally absorbed radiation successively passes through the surface and the rear barriers, thus providing the wavefield distribution between the barriers and the modulation of the distribution sections by changing the external voltage.

Technologically, such a structure can be implemented in the process of receiving the IC which is processing the output signal of the photoreceiver (as a monolithic solid-state IC). The thicknesses of the active layer d and the rear diffusion layer w should be

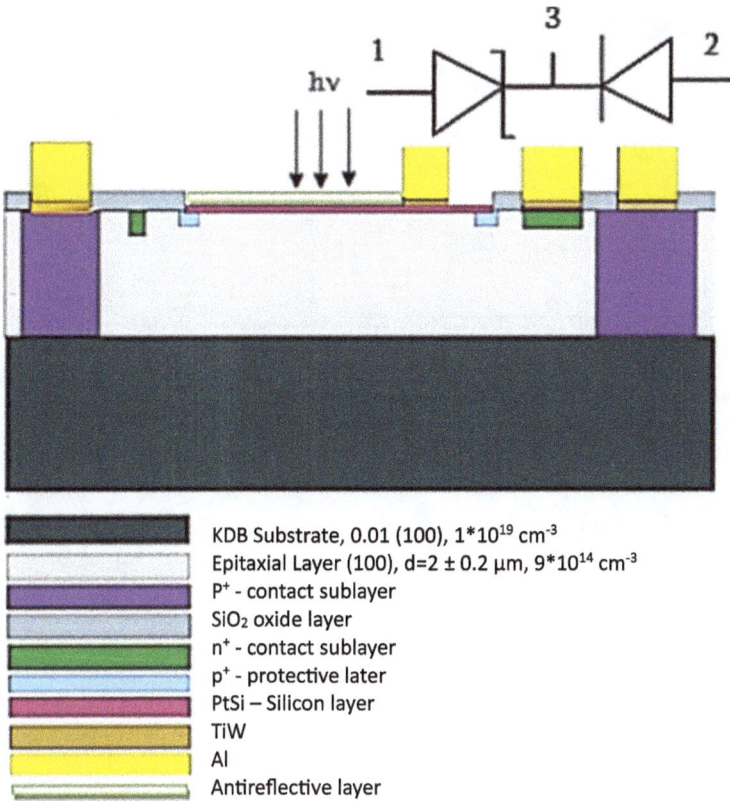

KDB Substrate, 0.01 (100), $1*10^{19}$ cm^{-3}
Epitaxial Layer (100), d=2 ± 0.2 µm, $9*10^{14}$ cm^{-3}
P$^+$ - contact sublayer
SiO$_2$ oxide layer
n$^+$ - contact sublayer
p$^+$ - protective later
PtSi – Silicon layer
TiW
Al
Antireflective layer

Figure 5.1: The cross section and the structural-technological parameters of the photodetector structure.

chosen based on the target problems since they determine the longitudinal absorption limit. If it is necessary to significantly reduce the absorption of the $\lambda = 0.45\,\mu$m and longer waves, then, taking into account the depth of absorption of electromagnetic waves in silicon, for shortwave (Figure 5.2) and longwave (Figure 5.3), the thickness $d \leq 1.5\,\mu$m should be taken.

A translucent metal-semiconductor-silicide barrier with a surface size of $1,450\,\mu$m is formed on the upper surface of the $d = \sim2\ \mu$m thick n-type layer (received by an epitaxial growth or an ion doping). The relatively thick membranes of the barrier, deposited from the boundary TiW, Al metal, provide a silicon ohmic contact of the output. The topological image of the experimental samples of the photodetector developed within the framework of our cooperation with "RD ALFA Microelectronics" is shown in Figure 5.4.

An ohmic contact is formed on the p-type surface. The contact width is $50\,\mu$m, and when the photocurrents are up to a milliampere, it ensures the output of the currents without heating and the welding of the output gold wires. An antireflective SiO$_2$ film with a thickness of $0.05\,\mu$m is grown on the photosensitive surface. The transmission capacity of the film is 95% in the UV range of the spectrum (S. Khudaverdyan, Kha-

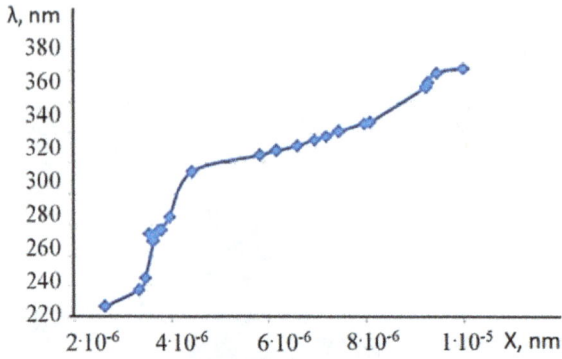

Figure 5.2: The dependence of short wavelengths on the absorption depth.

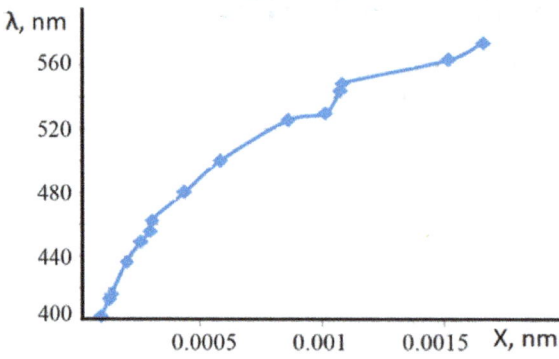

Figure 5.3: The dependence of long wavelengths on the absorption depth.

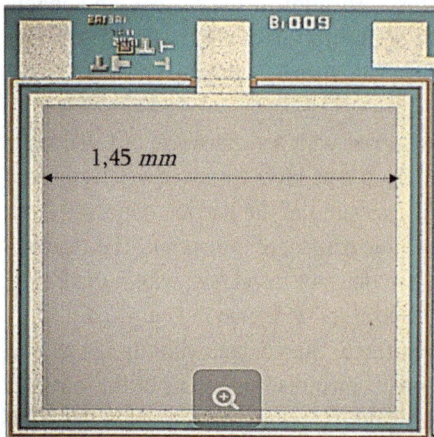

Figure 5.4: Photodetector topology.

chatryan, et al., 2013b; Mogniotte et al., 2018; Popov, 2003). At the same time, the thick SiO$_2$ layers of the surface, which are impermeable to the radiation, act as a shield. The logic of the spectral analysis carried out with the proposed spectrometer presumes the following:

- The creation of barriers with abrupt junctions to obtain small reverse dark currents, low noises, high threshold photosensitivity, operation speed, and temperature stability.
- The selection of the metal providing the highest possible value of the metal-semiconductor barrier, which will make it possible to ensure the significant size of the silicide potential barrier even at low values of the impurity density in the base.
- The attainment of such portions of impurity densities in the rear diffusion layer and in the base which will ensure high values of the rear potential barrier.
- The impurity density in the base that will ensure the contact of the oppositely directed potential barriers and the linear change of the widths of the barriers (one at the expense of another) by the external voltage, provided that the above-stated conditions are met. It will make it possible to register the individual waves from the longitudinally absorbed radiation with the help of an appropriate algorithm.
- A sealed housing is available for the photodetector, which must have a penetration window for beams of at least 200 nm in length to allow spectral analysis from 200 to 1,100 nm.
- Years of research have shown that the best silicon barrier can be formed with the PtSi system. Different sources indicate different heights of potential silicon barriers: 0.84 eV (Medevedev, 1970) and 0.82 eV (Komarov et al., 2011).

In further calculations, the value 0.84 eV will be used.

5.3 Relating structural, energy, and technological parameters of photodetectors

Let us consider the changes in the structural, energy, and technological parameters based on the changes in the impurity density in the base. The contact of the depleted layers of the surface and rear barriers, $(0 - x_m)$ and $(d - x_m)$ respectively, is provided in the base. Taking that the base is occupied with the depleted layers of the barriers (Figure 5.5), the potential distribution of x_m in it is determined by the field potential V and the space charge density N_d creating that field (I. Vikulin & Stafeev, 1990).

The following process takes place. Given that the base is occupied with the depleted layers of the oppositely directed barriers, it is possible to determine the potential distribution of the electrons in the conduction band. To do that, it is necessary to solve the Poisson equation. It connects the field potential to the volume density of the charges creating that field:

$$\frac{d^2V(x)}{dx^2} = -\frac{\rho}{\varepsilon\varepsilon_0} \tag{5.1}$$

In Poisson's equation, let us pass from the potential $V(x)$ to the potential energy of the electron $\varphi(x)$, $(\varphi(x) = -qV(x))$.

Since $\rho = qN_d$, we derive the following equation:

$$\frac{d^2\varphi}{dx^2} = \frac{q^2 N_d}{\varepsilon\varepsilon_0} \tag{5.2}$$

where N_d is the density of the donor impurity in the base, ε is the permittivity of the material, ε_0 is the permittivity of free space, and q is the electron charge.

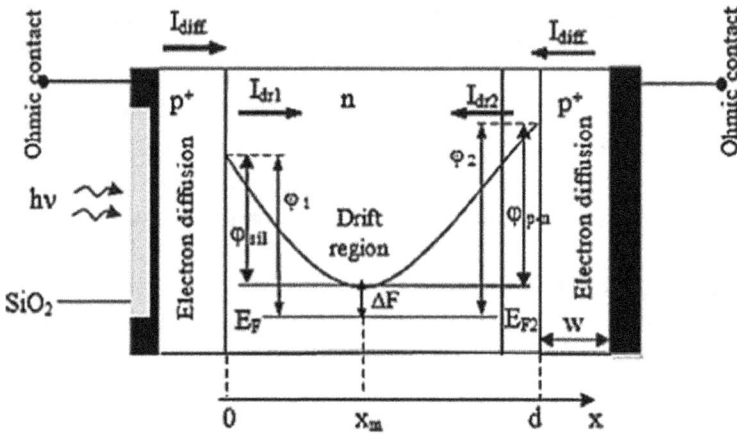

Figure 5.5: The p⁺-n-p⁺ structure, the distribution of the potential energy of the electrons in the conduction band, and the directions of the currents (reproduced with permission).

The boundary condition for this equation is $(|d\varphi/dx| = 0)$, when $x = x_m$ (Figure 5.5, x_m is the minimum of the potential energy of the electrons), and $\varphi(x) = \varphi_1$ when $x = 0$. Given the boundary conditions, let us integrate eq. (5.2). As a result, we will get

$$\varphi(x) = \frac{q^2 N_d}{2\varepsilon\varepsilon_0}x^2 - \frac{q^2 N_d}{\varepsilon\varepsilon_0}x \times x_m + \varphi_1 \tag{5.3}$$

When $x = d$ and the external voltage is applied (Figure 5.6):

$$\varphi(x) = \frac{q^2 N_d}{2\varepsilon\varepsilon_0}d^2 - \frac{q^2 N_d}{\varepsilon\varepsilon_0}d \times x_m + \varphi_1 = \varphi_2 + qV \tag{5.4}$$

Thus, it is possible to determine the dependence of x_m and $d - x_m$ on the external voltage

$$d - x_m = \frac{d}{2} + \frac{\varepsilon \varepsilon_0 (\Delta \varphi + qV)}{q^2 N_d d}$$

or

$$x_m = \frac{d}{2} - \frac{\varepsilon \varepsilon_0 (\Delta \varphi + qV)}{q^2 N_d d} \tag{5.5}$$

where $\Delta \varphi = \varphi_2 - \varphi_1$ or $\Delta \varphi = \varphi_{sil} - \varphi_{p-n}$.

If the two barriers have the same height, then $\varphi_2 - \varphi_1 = 0$. Thus we get

$$d - x_m = \frac{d}{2} + \frac{\varepsilon \varepsilon_0 qV}{q^2 N_d d}$$

or

$$x_m = \frac{d}{2} - \frac{\varepsilon \varepsilon_0 qV}{q^2 N_d d} \tag{5.6}$$

With the help of (5.4), we can determine the modulation depth of the layers depleted by the external voltage and the position of x_m. With the help of $\Delta x_m = x_{m2} - x_{m1}$, we can determine the length and the intensity of the wave.

With the help of (5.1) and (5.2), as well as the expression for the position of the Fermi level $\Delta F = (KT \ln N_c)/n_n = 0.27$ eV (Khudaverdyan S. Kh et al., 2017, 2013), the dependences $x_m(V), x_m(N_d), \varphi_{p-n}(N_d), \varphi_{sil}(N_d), d(N_d), \Delta F(N_d)$ were studied (Figures 5.7–5.13), and the structural parameters were determined under the conditions of the contact of the potential barriers in the point x_m.

When the impurity density is increased, the distance of the Fermi level from the bottom of the conduction band decreases faster at low densities than at high densities (Figure 5.7). The height of the silicon barrier created by Pt with n-Si is 0.84 eV. The difference between that height and ΔF in the base, at different impurity densities, will give the dependence of the height of the silicon barrier on the impurity density (Figure 5.8):

$$\varphi_{sil} = \varphi_1 - \Delta F = \varphi_1 - KT \ln(N_c/n_n) = 0.84 - 0.27 = 0.57 \text{ eV} \tag{5.7}$$

The height of the potential barrier of the p-n junction is determined by the following expression (Sze et al., 2021):

$$\varphi_{p-n} = \frac{KT}{q} \ln \frac{p_p \times n_n}{n_i^2} = 0.76 \text{ eV} \tag{5.8}$$

It has a constant value of 0.19 eV. Under the condition of the contact of the potential barriers in the base, with the help of the expressions determining the widths of individual barriers,

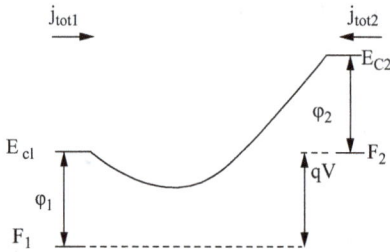

Figure 5.6: The change in the energy of the conduction band under the influence of the external voltage, and it increases according to the logarithmic law depending on the impurity density in the base (Figure 5.8). In (5.8), kT/q, the heat potential, is 0.26 eV at 300 K, n_n is the electron density in the base, p_p is the hole density in the rear p$^+$ region, and n_i, the density of free charges in the pure semiconductor, is 1.6×10^{10} cm^{-3} for Si at 300 K.

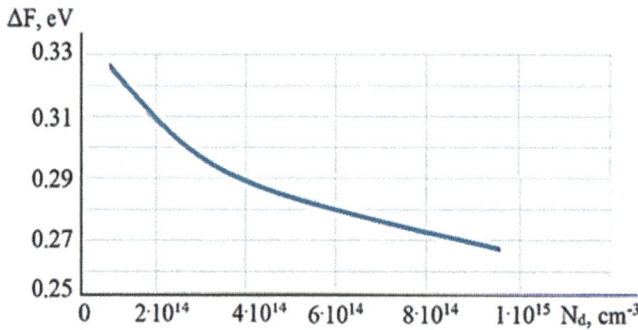

Figure 5.7: The dependence of the distance of the Fermi level from the bottom of the conductor band on the impurity density.

$$d_{\text{p-n}} = \sqrt{\frac{2\varepsilon\varepsilon_0\varphi_1}{q^2 N_d}}, \quad d_{\text{PtSi}} = \sqrt{\frac{2\varepsilon\varepsilon_0\varphi_2}{q^2 N_d}} \qquad (5.9)$$

we will obtain the dependence of the base width $d = d_{\text{p-n}} + d_{\text{PtSi}}$ on the impurity density. Thus, by selecting the heights of the oppositely directed potential barriers (Figures 5.8 and 5.9), we can determine the width of the base as a technological parameter (Figure 5.10).

The algorithm of the selection of the individual waves from the integral flux of the radiation presumes the absorption depth of a given wave of the order of x_m. Therefore, the change of x_m depending on the external bias voltage (at the constant impurity density, Figure 5.11) and the impurity density in the base (at the constant voltage, Figure 5.12) is important.

As shown in Figures 5.11 and 5.12, when the voltage values are negative, the surface potential barrier is reverse biased, and x_m increases with the increase in the negative

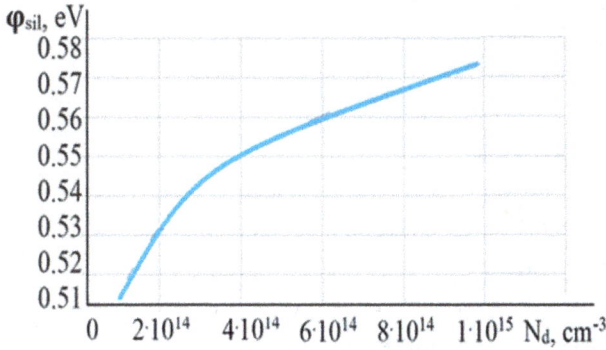

Figure 5.8: The dependence of the height of the potential energy created by PtSi on the impurity density.

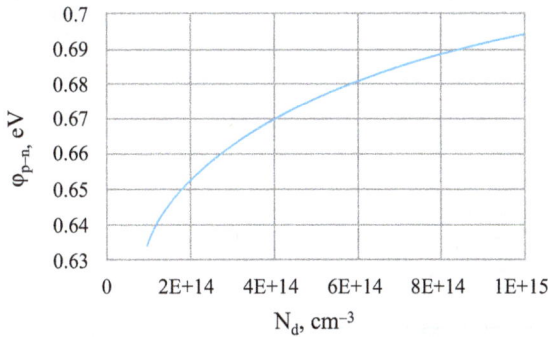

Figure 5.9: The dependence of the height of the potential energy of the p-n junction on the impurity density.

voltage. In Figure 5.11, the dependence is linear. It is evident that the slope of the curves is bigger and the change in the barrier widths (one at the expense of another) is more active at the relatively smaller values of the impurity densities in the base.

Thus, with an increase in the impurity density in the base, the silicon potential barrier and the potential barrier of the p-n junction increase at low densities first rapidly, and then slowly, tending to the saturation value of 0.58 eV in the case of the first barrier and of 0.7 eV in case of the second barrier, within the investigated density range.

As opposed to that, the base width, depending on the external voltage, decreases first rapidly and then slowly, tending to the saturation value of 0.00008 cm:

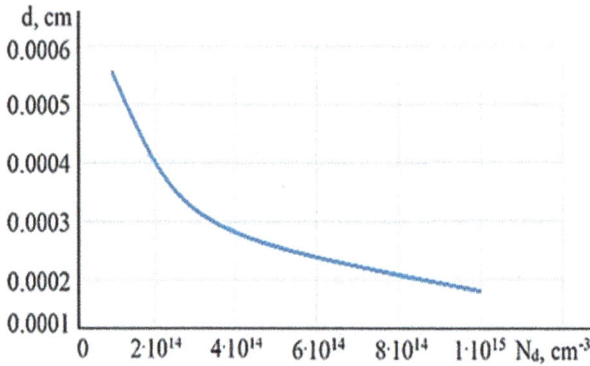

Figure 5.10: The dependence of the base width on the impurity density.

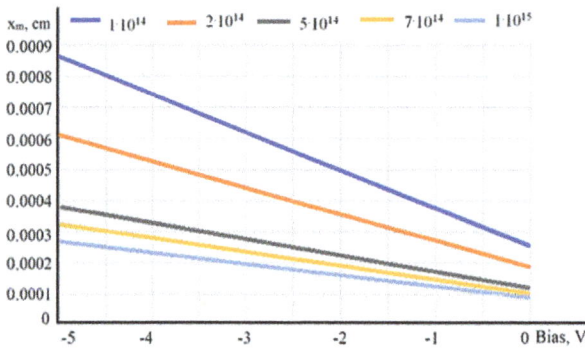

Figure 5.11: The dependence of the point of minimum on the potential energy of the electron in the base on the external voltage at different impurity densities.

–   Under the same conditions, the contact point x_m of the barriers (Figure 5.5) decreases linearly when the rear barrier is reverse biased and the surface barrier is forward biased. The smaller the impurity density in the base, the faster the decrease (Figure 5.11).
–   In the absence of the bias voltage, the smaller the impurity concentrations in the base, the larger the widths of the oppositely directed potential barriers, under the conditions of their contact, and the larger the width of the base making their sum.

Figure 5.13 presents the dependence of the residual intensity of the absorbed radiation on the wavelength in the point x_m at base widths of 3.5 and 1.6 μm.

The spectral dependence of the solar radiation intensity is used. As shown in the figure, when the base width $d = 1.6\,\mu m$ in x_m (Figure 5.2), the waves with a length of up to 400 nm are mostly absorbed (the residual intensity is low). In the case of larger values of x_m (e.g., when $d = 3.5\,\mu m$, the longer waves penetrate the absorption range (Figure 5.13).

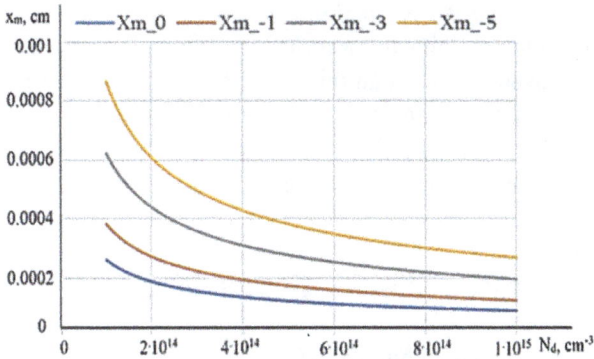

Figure 5.12: The dependence of the point of minimum on the potential energy of the electron in the base on the impurity density at different external voltages.

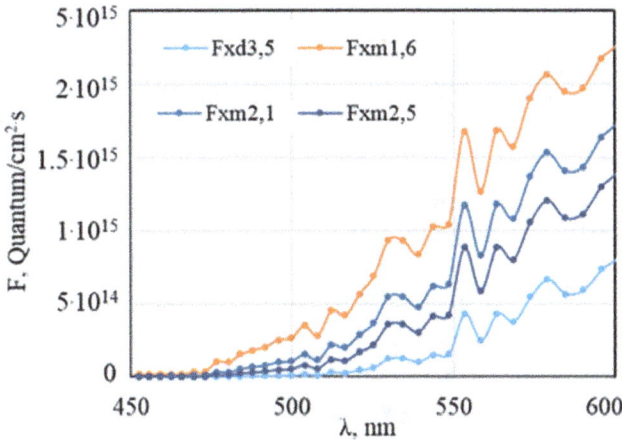

Figure 5.13: The dependence of the residual intensity of the absorbed radiation on the wavelength in the point x_m at base widths of 3.5 and 1.6 μm.

The mechanism of the mentioned radiation absorption and the structural regularities of the photodetector show that the larger the x_m, that is, the larger the base, the closer the residual intensity wave that reaches x_m to the single value. That is, to obtain high-precision spectral selectivity, it is necessary to have a large base. To obtain a large base, it is necessary to carry out long-term, high-temperature deposition of the epitaxial layer. In the process, the self-diffusion from the rear p^+ layer may occur, leading to a decrease in the potential barrier abruptness, and thus reducing the accuracy of the selective sensitivity.

Therefore, the shorter the term and the lower the temperature of the epitaxial growth, the higher the accuracy of determining the wavelengths and their intensities. In that case, the base will be relatively narrow, and the number of residual intensity waves that reach x_m may be more than one. It is also necessary to obtain low-temperature silicide. Hence, to obtain the samples, the technological processes described in Komarov et al. (2011) and Shalimova (1985) are used.

5.4 Structural layers of the spectrometric photodetector

The technological processes are considered within the technological cycle of obtaining integrated circuits (Chernyaev, 1977; Koledov, 1989; Shalimova, 1985).

5.4.1 Forming p⁺ and n layers

First, using photolithography and chemical etching, the windows are opened on the SiO_2 surface and deposited on a relatively high-ohmic p-type conductivity platform. In the following processing steps, the pockets are etched. Through the epitaxial growth, a p⁺-type conductivity intercontact region is formed in the pockets. The growth starts from the gas phase containing boron, which provides the p⁺-type conductivity of the obtained membrane. The chemical reaction (CR) of the growth is given as follows:

$$2BCl_3 + 3CH_2 \rightarrow 2P + 6\,HCl \tag{CR1}$$

Then, again using chemical etching, a 2-μm-thick single-crystal base is formed by the epitaxial growth in the pockets opened in appropriate places (Figure 5.1). In this case also, the process starts from the gas phase during which the following reaction provides the n-type conductivity of the obtained membrane utilizing the imported phosphorus

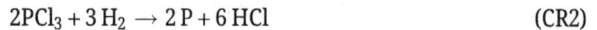

$$2PCl_3 + 3\,H_2 \rightarrow 2\,P + 6\,HCl \tag{CR2}$$

The deposition takes place at a temperature of about $1,150\ °C$, which ensures the repetition of the crystal lattice structure of the platform in the growing membrane, that is, the epitaxial growth of the membrane. The growth rate is $\sim 0.5\,\mu m/min$. The technology described above can provides the required mixture density in the membrane. This should provide the base with the required thickness occupied by the oppositely directed potential barriers and the effective modulation of its conductivity. The high temperature growth of the membrane results in self-diffusion between the platform and the base, which affects the abruptness of he barrier junction.

It should be noted that the technology used for obtaining an epitaxial membrane with the required parameters makes it possible to ensure the high accuracy of the thermal process, the required purity, homogeneity, and composition of the gas-vapor mixture, the required gas-dynamic state near the platform surface, the perfection and purity of the platform structure, and the required growth rate of the membrane.

It should also be noted that the vacuum technology of the epitaxial growth of the membrane currently used in nanotechnologies (molecular beam epitaxy, MBE) ensures the growth of the epitaxial layers with the accuracy of the atomic layers on the surface of the high-purity silicon platform at the temperature of 300–500 °C in the conditions of high vacuum $(10^{-10}-10^{-11}$ Pa). To ensure the purity of the working volume, the platform is heated by a high-frequency field or an electronic beam. In this process, the silicon atoms get diffused over the platform surface toward the points on the crystal lattice where they have minimum free energy, thus forming an epitaxial membrane. However, this technology cannot provide the sufficient base thickness required for effective radiation absorption, since the process can take longer time.

5.5 Ohmic contacts and the technological parameters of the structure

The ohmic contacts are formed in the windows opened on the surface of the p^+ region using the vacuum deposition, the successive deposition of tungsten and aluminum membrane, and the 5 min thermal treatment in vacuum at a temperature of 400 °C. The silicon plate is subjected to technical and chemical purification before each technological process. However, the results of the experiment show that the high-temperature epitaxial growth of the membrane results in the self-diffusion of the impurities between the platform and the developing base, which affects the abruptness of the p-n junction. The latter, in turn, increases the dark currents and the noise and reduces the photosensitivity and the effective conductivity modulation of the base required for the spectral-selective sensitivity. Therefore, it is necessary to have an optimal technology that ensures the minimum duration of the high-temperature processes. For that, the method of the ion-implantation doping of the conductivity layers has been tested.

A technically feasible structure has been developed in collaboration with RD Alfa Microelectronics. Several options for the samples have been produced and tested. Figure 5.1 shows one of the most acceptable designs. To create the structure, the dose of the ion-implantation doping (Q), the temperature of the process (T), and the duration of the process required for forming each layer (t) have been used. Figure 5.14(a) presents the topological image of the detector. The detector is placed in the case with a transparent window display. The case ensures the hermeticity of the structure (Figure 5.14c, d). In Figure 5.14b, the withdrawal of the sample into the housing is indicated. Outputs 1, 2, and 8 of the case are used: 1 comes out of the silicon contact, 2 comes out of the p^+ contact of the rear junction, and 8 comes out of the base contact. A 1-mm-thick glass plate produced by the German company SCHOTT is used as the case window. It has the spectral dependence of the beam transmission presented in Figure 5.15. Later,

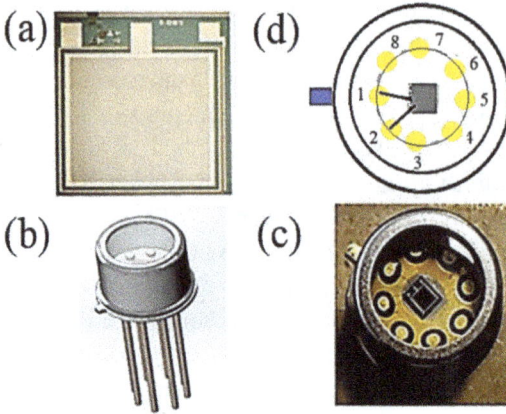

Figure 5.14: (a): The topological image of the photodetector spectrometer, the cross section of the case, the top view, and the side view (b–d).

by replacing the window glass with the quartz glass, it will be possible to obtain the UV radiation transmission starting from 200 nm.

Figure 5.16 presents the spectral dependence of the reflection index of the glass. As shown in the figure, the radiation transmission starts at the wavelength of 300 nm. About 90% of the radiation transmission starts from the wavelength of 350 nm. It meets the requirements of our experiments since quite high powers of the incident radiation are used: the distance between the LED and the photodetector is not more than 2–3 cm.

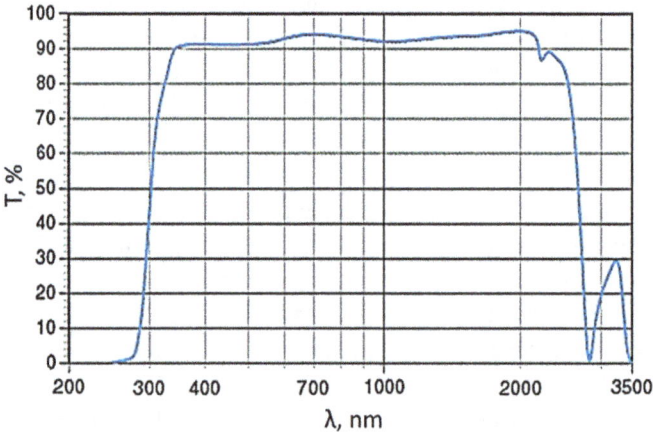

Figure 5.15: The spectral dependence of the transmission of the case window.

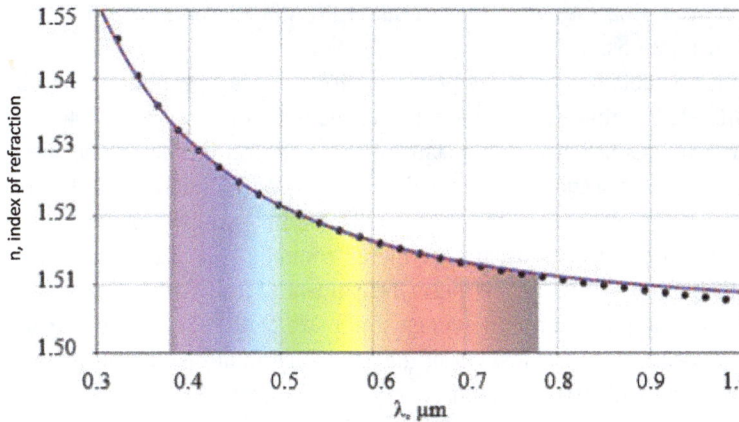

Figure 5.16: The spectral dependence of the reflection index of the window.

5.6 Choice of the silicide barrier

The choice of metal contact is very important since it has the greatest influence on the electrical properties of the semiconductor structure with the surface barrier and determines the height of the silicide barrier. In the case of lower barriers, the silicon diode has a rather high density of reverse currents, which also increases by the exponential law, depending on the reverse bias voltage and the temperature. The reverse current can be reduced if a metal with a high barrier compared to silicon is taken (Komarov et al., 2011; Sze et al., 2021).

The silicides of transition metals play an extremely important role in the production of Schottky diodes. The silicides of metals have high conductivity close to the metal conductivity, high-temperature stability, and a high potential barrier with silicon. One of the most popular substances in this class is the silicide of platinum. It makes it possible to obtain structures with small reverse currents, high breakdown voltage, and high operating temperatures up to 175–200 °C.

The standard technological mode of the formation of the silicide of platinum is the thermal treatment at 550 °C for 15–30 min. The time of the treatment is determined by the thickness of the output metal layer. The main disadvantage of this mode is the formation of a nonhomogeneous membrane.

It is known that the silicide of platinum can also be obtained at lower temperatures (Komarov et al., 2011). Thus, it will be possible to obtain a more homogeneous silicon layer by lowering the temperature and extending the time, since the formation of silicide is more uniform. Moreover, the Pt_2Si phase is formed at low temperatures, and the PtSi phase is formed at relatively high temperatures. In the investigated structures, the low-temperature range of 100–300 °C is used to form the silicide of platinum. The technological processes were carried out at RD Alfa Microelectronics by undertaking joint consultations.

The metal film was obtained from the platinum target by magnetron deposition in the vacuum of ~10^5 Pa. The thickness of the film changed within the range of 7–40 nm. The thermal treatment was performed in a standard technological furnace at the specified temperatures. The membranes were studied under an electron microscope. The results are presented in the pictures. The output layer has a fine-grained polycrystalline structure. As a result of the thermal treatment, the individual grains of the Pt$_2$Si phase were formed starting from 200 °C. At the thermal treatment of 240 °C, the entire metal layer consists of the Pt$_2$Si phase with an average grain size of 40–80 nm. The thermal treatment at 260 °C results in the transition of the entire platinum membrane into the Pt$_2$Si silicide with clear boundaries between the crystals. The small PtSi silicide crystals are formed when the temperature is above 260 °C. With the increase in the temperature, the grain size decreases, and at 360 °C, the PtSi silicide phase is completely formed.

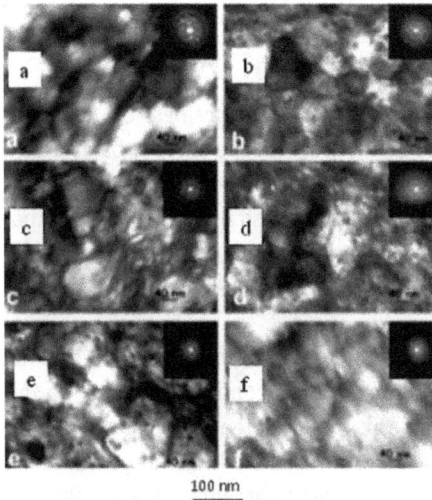

100 nm

Figure 5.17: TEM (transmission electron microscope) – micrograph and microdiffraction images. The metal film was obtained from the platinum target by magnetron deposition in the vacuum of ~10^5 Pa. The thickness of the film changed within the range of 7 – 40 nm. Pt/Si structure after the deposition and thermal treatment: (a) after the deposition; (b) at 200 °C, 480 min; (c) at 220 °C, 480 min; (d) at 240 °C, 240 min; (e) at 260 °C, 240 min; and (f) at 280 °C, 240 min.

Thus, it is possible to obtain Pt$_2$Si and PtSi silicides by low-temperature processing of the platinum film deposited on the silicon platform. The formation of silicide begins at a temperature of 180 °C with the formation of Pt$_2$Si phase at the Pt/Si boundary. The thermal treatment at temperatures of 240–260 °C results in the complete transition of the platinum film into the structurally homogeneous Pt$_2$Si silicide, as shown by TEM micrographs in Figure 5.17 (a-f). The transition into the PtSi phase is observed after the thermal treatment at a temperate of 360 °C.

The experimental samples of the cased photodetector have been studied. The $d = 2\,\mu m$ thick base is surrounded by the barriers of the silicide and the p-n junction from the opposite sides. When the specified thickness of the base is provided, the growing process that takes place from the platform to the base by self-diffusion is relatively short (about 4 min), and the sharpness of the p-n junction and the specified width of the base are not violated. The outputs of the structure are connected to the outputs 1, 8, and 2 in this case (Figure 5.1). As has already been mentioned, the n-base of the structure is a phosphorus-doped epitaxial layer with the mixture density of $N_d = 9 \times 10^{14}$ cm^{-3}. The p$^+$ platform is boron-doped with the mixture density $N_a = 1 \times 10^{19}$ cm^{-3}. These parameters provide the required contact of the oppositely directed barriers in the base. The mixtures are ionized at room temperature and determine the densities of the majority charge carriers n_n and p_p within the specified range. The density of the self-charges of silicon is $n_i = 1.6 \times 10^{10}$ cm^{-3}. and the thermal energy at room temperature is 0.026 eV (Komarov et al., 2011; I. Vikulin & Stafeev, 1990).

The distance between the base of the conduction band and the Fermi level in the base is $\Delta F = 0.27$ eV, and the height of the silicide Schottky barrier is $\varphi_1 = 0.84$ eV (Sze et al., 2021). Therefore, the height of the PtSi silicide barrier created by Pt is $\varphi_{sil} = 0.84 - 0.27 = 0.57$ eV. The height of the potential barrier of the p-n junction in (5.8) is $\varphi_{p-n} = 0.76$ eV. Thus, the difference between the oppositely directed potential barriers is 0.76–0.57 = 0.19 eV.

Based on the data mentioned above, that is, the permittivity of silicon and free space $\varepsilon = 12$ and $\varepsilon_0 = 8.86 \times 10^{14}$ F/cm, respectively, the impurity density in the base and the electron charge, and taking into account that the base resistance is very high as compared to the boundary layers, it can be admitted that the space charges are placed in the base and their widths can be determined in the following way:

$$d_{p-n} = \sqrt{\frac{2\varepsilon\varepsilon_0\varphi_2}{q^2 N_d}} = 0.917\,\mu m, \quad d_{PtSi} = \sqrt{\frac{2\varepsilon\varepsilon_0\varphi_1}{q^2 N_d}} \sim 1\,\mu m \qquad (5.10)$$

The total width of the potential barriers of the structure is 1.917 μm. Therefore, the base $d = 2 \pm 0.2\,\mu m$ is almost entirely occupied by the oppositely directed potential barriers. The external voltage mainly falls on the potential barriers in the base, and the position of their contact point can be controlled by the external voltage.

5.7 Summary

- The regularities of the shift of the contact point of the potential barriers in the base have been studied, based on the impurity density and the applied external voltage. It has proved to be the most effective at low voltages and low impurity densities.

- The point of contact x_m of the barriers changes linearly under the influence of the external voltage. Moreover, the lower the impurity density in the base, the faster the change.

- The lower the impurity density in the base, the wider the oppositely directed potential barriers and the larger the base consisting of their total width.

- For high-precision spectral selectivity, it is necessary to have a wide base for the registration of longwaves and a narrow base for the registration of shortwaves. However, the self-diffusion from the rear p^+ layer, which reduces the sharpness of the potential barrier, restricts the deposition time of the epitaxial layer and the possibility of obtaining a wide base.

- The best experimental parameters of the modulation of the heights of the potential barriers and their widths are provided by $p^+(PtSi)$-$n(Si)$-$p^+(Si)$ structures with a base width of $2\,\mu m$ and the heights of the oppositely directed barriers 0.76 and 0.57 eV.

- Pt has been chosen as the silicide metal, since, compared to the other metals, it provides low-temperature treatment, a large potential barrier of the silicide PtSi with silicon, and therefore, the most effective modulation of the barrier width.

Part 4: **Measurement methods and photoelectronic
characteristics**

Chapter 6
Photoelectronic characteristics and measurement methods

6.1 Introduction

Photoelectronic characteristics refer to the properties exhibited by materials or devices when subjected to light, particularly in terms of their interaction with photons. These characteristics are crucial in various fields such as optoelectronics, photovoltaics, and photonics. Understanding and measuring these characteristics are essential for designing and optimizing devices for specific applications. Some of the key photoelectronic characteristics and measurement methods are photoelectric effect, quantum efficiency, spectral response, responsivity, dark current: noise characteristics, and shot noise. Understanding and characterizing noise is crucial for assessing the overall performance and signal quality of a device. Typical measurement methods employed to characterize these parameters are spectrophotometry, current-voltage (UV) characteristics, quantum efficiency, noise measurement, and transient response. These measurements provide insights into factors such as carrier lifetime and response time. In general, a thorough understanding of photoelectronic characteristics and appropriate measurement methods is essential for the design, optimization, and performance evaluation of optoelectronic devices and systems.

6.2 Current-voltage characteristics

The current-voltage characteristics were measured by high-precision measuring instruments Ketlin 6340 and B2912A. The voltage penetrability was 1 μV within the voltage range of ± 2 V with the accuracy of $\pm 0.02\%$. The current penetrability was ± 1 pA within the range of ± 100 μA with an accuracy of $\pm 0.02\%$ The voltage changed from +5 to -5 V with the step of 1 mV. The output spectrum was restored with the help of the numerical data of the current-voltage characteristics and the theoretical calculations. In the investigated photodetector, the dark currents are several orders of magnitude smaller than the light currents depending on the length and the intensity of the used wave. Therefore, the above-mentioned accuracies are sufficient for the research.

The measuring instruments show an output to the computer, thus providing the digital data of the current-voltage characteristics in the automatic operation mode. The schematic solutions eliminate the unwanted voltage deviations and reduce the inaccuracies in the current-voltage characteristics of the photodetector during the

https://doi.org/10.1515/9783111428024-007

generation of the step linear voltage. The resulting current-voltage characteristics have clearly defined positive and negative branches (Figure 6.1).

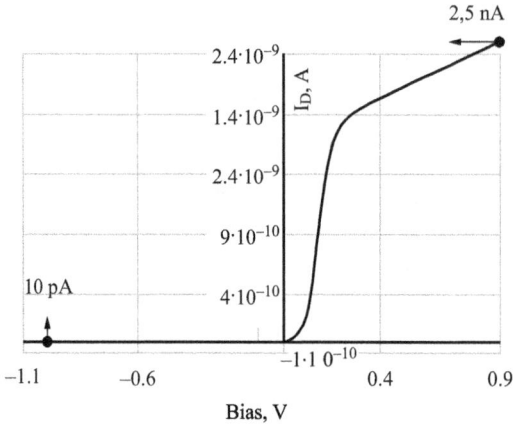

Figure 6.1: The dark current-voltage characteristics of the photodetector taken by the measuring instruments.

The zero current is obtained at a certain positive value of the voltage, that is, the value equivalent to the difference in the heights of the oppositely directed potential barriers. When the impurity density in the base for the investigated sample is 9×10^{14} cm^{-3}, the measured value is 170 mV and the calculated value is 190 mV, that is, the values are close to each other.

Figure 6.2 presents the circuit connection of the photodetector. Contact point 8, as shown, is accessible out of the n-base, thus making it possible to individually study the photoelectric properties of the oppositely directed potential barriers p$^+$-n and n-p$^+$ with the help of one or two contacts. The measurement circuit of the current-voltage characteristics supposes the simultaneous measurement of the voltage/current values of both barriers with the corresponding measurement devices.

Figures 6.3 and 6.4 show the current-voltage characteristics of individual barriers at the longitudinal absorption of the radiation incident from the side of the silicide barrier.

The rear n-p$^+$ junction with a higher barrier ensures a larger value of the saturation photocurrent and a smaller change (Figure 6.3) than the p$^+$-n junction (Figure 6.4), which is due to the more abrupt rear junction, the large value of the absorption depth of the wave $\lambda = 705$ nm, and the presence of the diffusion component of the photocurrent in the rear p$^+$. Figure 6.5 shows the current-voltage characteristics of the p$^+$-n-p$^+$ structure under the conditions of the absorption of the same wave when the voltage is applied to contacts 1 and 2. The change in the sign of the voltage value brings to the photocurrent saturation of the oppositely directed potential barriers, each within the range of its reverse bias.

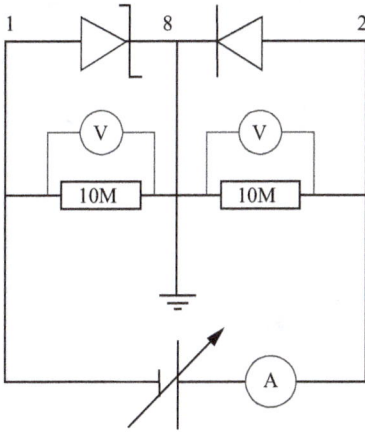

Figure 6.2: The measurement circuit of the current-voltage characteristics of individual diodes.

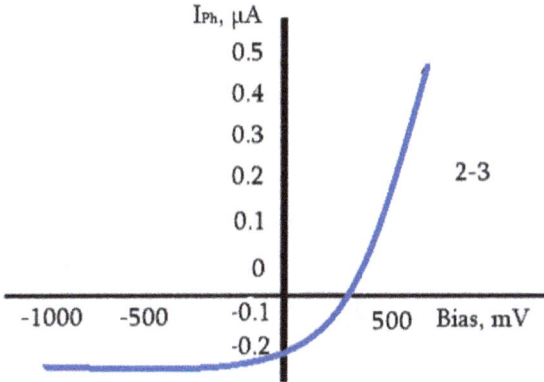

Figure 6.3: The current-voltage characteristics of the n–p$^+$ barrier at the wave absorption $\lambda = 705$ nm.

At negative voltages, the photocurrent is conditioned by the saturation currents of the rear n-p$^+$ junction, while at positive voltages, the photocurrent is conditioned by the saturation currents of the near-surface p$^+$-n junction. In the latter case, the saturation progresses smoothly (Figure 6.5, positive voltages), which is probably the result of the counterreaction of the higher rear potential barrier conditioned by the increase in the applied voltage. In the first case, the lower surface potential barrier is small; it causes insignificant opposition and a sharp saturation of the photocurrent (Figure 6.5, negative voltages).

In the investigated p$^+$-n-p$^+$ structure, at the longitudinal absorption of the radiation, depending on the wavelength, different proportions of the photocurrents of the oppositely directed potential barriers are observed. The reciprocal of the absorption coefficient of the indicated wavelength $1/\alpha = 2.6\ \mu m$ defines the absorption depth. The

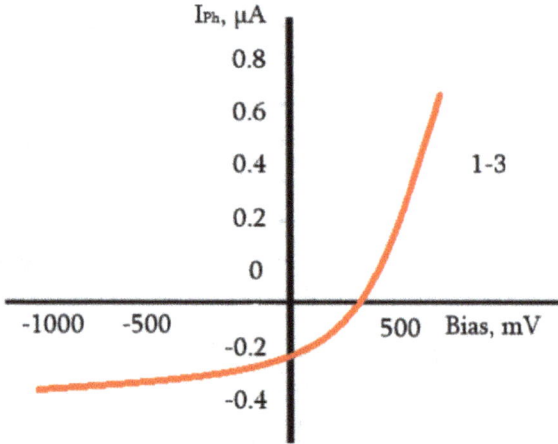

Figure 6.4: The current-voltage characteristics of the p^+-n barrier at the wave absorption $\lambda = 705$ nm.

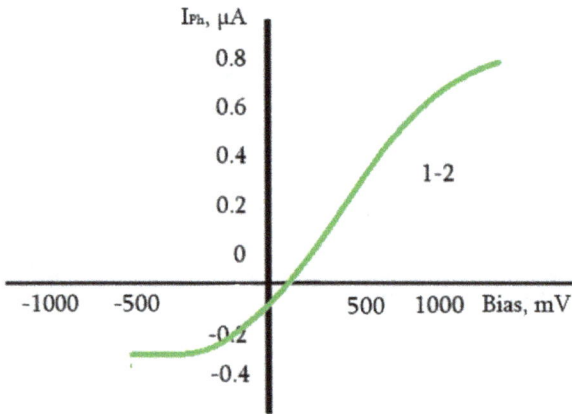

Figure 6.5: The current-voltage characteristics of the p^+-n-p^+ structure at the wave absorption $\lambda = 705$ nm.

width of the n-base of the investigated samples is $2\,\mu m$. As mentioned earlier, the rear barrier penetrates the base to a depth of $\sim1\,\mu m$.

In that case, taking into account the diffusion p^+ range, the wave will be largely absorbed within the active range of the rear barrier and will generate a photocurrent that is greater than the photocurrent of the near-surface barrier. Similar regularities are observed when using the measurement circuits in Figures 6.6–6.8. When the potential barriers are studied separately, the abrupt saturation of the opposite photocurrents is observed for both junctions.

Both the dark and light saturation photocurrents of the rear barrier are smaller than those of the Schottky barrier. It is due to the greater height of the rear barrier in

0.66 mA

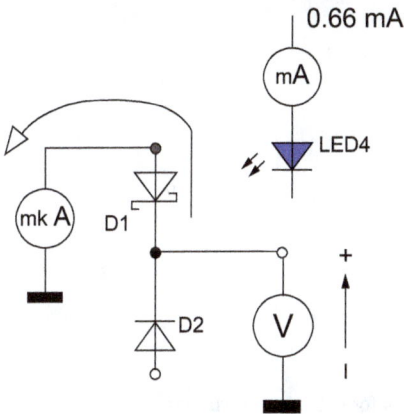

Figure 6.6: The circuit for obtaining the current-voltage characteristics of the Schottky barrier.

0.66 mA

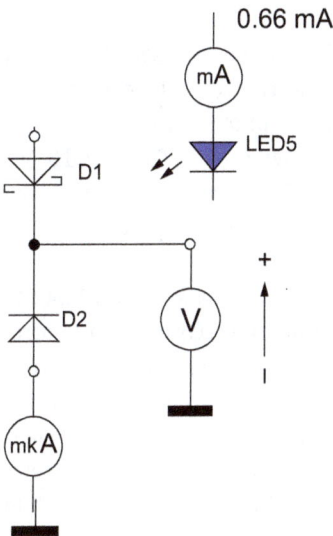

Figure 6.7: The circuit for obtaining the current-voltage characteristics of the rear barrier.

the case of dark photocurrent, and the homogeneity of the radiation absorption in depth in the case of light photocurrent.

Since the number of the absorbed quanta for all waves decreases in depth, the photocurrent generated by the near-surface barrier with the appropriate width will be greater than the photocurrent of the rear barrier. The absorption depth of the shortwaves is smaller than that of the longwaves. Therefore, in the investigated p^+-n-p^+ structures, at the longitudinal absorption of the radiation, different proportions of the photocurrents of the oppositely directed barriers are generated depending on the wavelength.

In that case, the shortwaves have a greater contribution to the photocurrent of the near-surface p^+-n barrier than to the photocurrent of the rear barrier. In the case

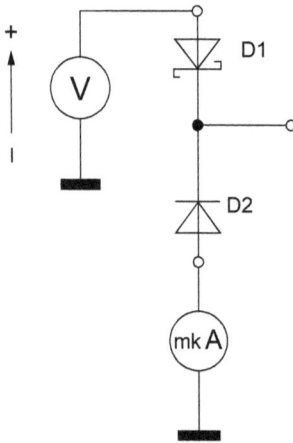

Figure 6.8: The circuit for obtaining the current-voltage characteristics of the p⁺-n-p⁺ structure.

of long waves, depending on the conditions, the situation can be reversed. Thus, with the increase in the external bias voltage, when the rear barrier gets wider in the base at the expense of the near-surface barrier, the shortwave photocurrent changes its sign (Figure 6.9) at a smaller value of the absolute voltage (at $\lambda = 462$ nm and 0.7 V) than the longwave photocurrent (at $\lambda = 660$ nm and 0.88 V). The current-voltage characteristics of the p⁺-n-p⁺ structure in Figures 6.3–6.5 were obtained by Ketlin 6340 at the wave absorption $\lambda = 705$ nm.

For further experiments, L-813SRC-J14 (AlGaInP), 153GC (GaP), and LL-304B-B4-GD (InGaN) LEDs were used as the radiation source. Their technical spectral maxima correspond to the wavelengths $\lambda_{max} = 660$ nm, $\lambda_{max} = 565$ nm, and $\lambda_{max} = 462$ nm, respectively. Figure 6.10 shows that the dark saturation current is of the order of several tens of pA, and is about 6 orders of magnitude smaller than the light saturation current. That makes it possible to register weak signals and have low noises. Figure 6.11 shows the current-voltage characteristics for different light sources, such as blue, green, red, and white LEDs and xenon lamps.

The saturation of the negative voltages concerns the rear barrier. Since the radiation sources have different intensities, it is impossible to determine the regularity of the current change from the wave absorption. It can only be said that all the curves have zero current at a certain positive value of the voltage. As mentioned earlier, this corresponds to the 0.19-mV difference of the potential barriers.

To reveal the above-mentioned regularities, the current-voltage characteristics of the photodetector with the same number were obtained at equal light intensities (Figure 6.12). In this case, the saturation of negative voltages concerns the near-surface barrier. It can be seen that the saturation photocurrent generated by the shortwave (blue light) in the near-surface barrier is greater than the photocurrent generated by the longwave (green and red lights).

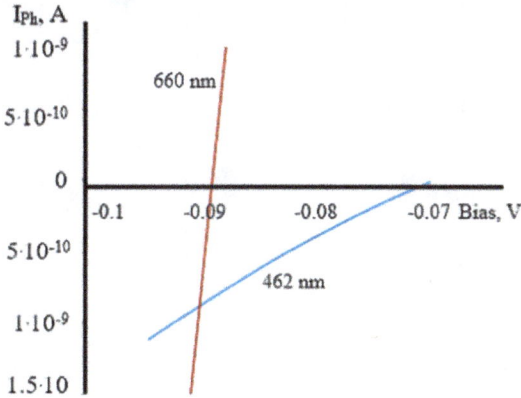

Figure 6.9: The change in the photocurrent sign at different wavelengths depends on the bias voltage.

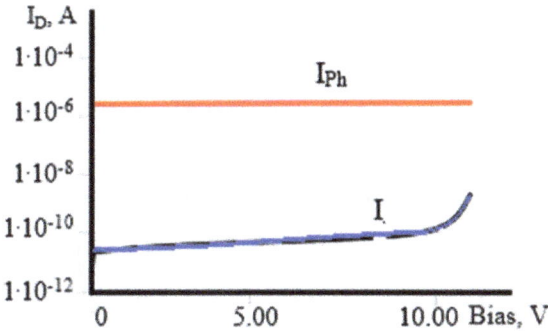

Figure 6.10: The current-voltage characteristics of the photodetector in the dark and under light exposure, on the semilogarithmic scale, when the rear barrier is reverse-biased.

It is because the shortwave quanta of the light are mostly absorbed near the surface and generate more charge carriers, and hence the photocurrent than the longwave quanta.

6.3 Volt-farad characteristic

The capacitance of the flat capacitor $C = (\varepsilon\varepsilon_0 S)/d$ for the near-surface and rear barriers of the investigated photodetector will be $C_1 = (\varepsilon\varepsilon_0 S)/x_m$ and $C_2 = (\varepsilon\varepsilon_0 S)/(d - x_m)$, respectively. Taking into account the fact that the capacitances in the photodetector

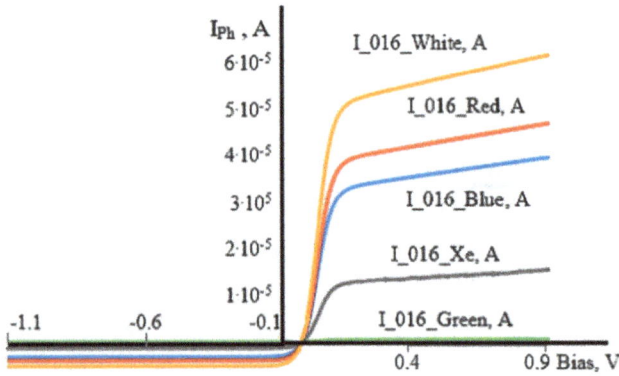

Figure 6.11: The current-voltage characteristics of the photodetector at the radiation absorption of blue, green, red, and white LEDs and xenon lamps with different intensities.

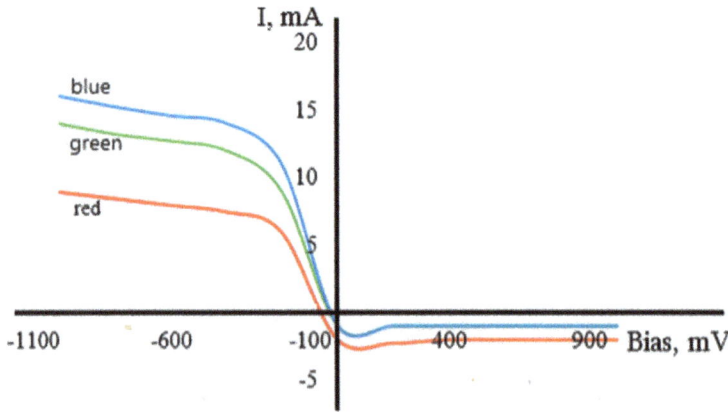

Figure 6.12: The current-voltage characteristics of the photodetector at the radiation absorption of the blue, green, and red LEDs with equal intensities.

structure are connected in series and have the general capacitance $C = C_1C_2/(C_1 + C_2)$, the expression for the base width will have

$$C = \varepsilon\varepsilon_0 S \frac{q^2 N_d d}{2 x_m q^2 N_d d + \varepsilon\varepsilon_0(\Delta\varphi + qV)} \tag{6.1}$$

The experimental results show that an increase in the voltage widens an individual potential barrier and reduces the capacitance. C–V characteristics are obtained using the measuring device immittance meter E7-25. The characteristics are taken in the dark (the black curve) and under the light exposure (the red curve). Figure 6.13 shows that the capacitance reduces with an increase in the reverse bias voltage of the Schottky barrier. Moreover, the registered capacitances are greater under light exposure than in the dark. It is obvious, since the light absorption reduces the height of the potential barrier, and hence the width d, which increases the capacitance. The same regularity is observed in the oppositely connected rear barrier (Figure 6.14). However, as compared to the Schottky barrier, the capacitance values in this case are smaller than in the case when the barrier height and the width are great.

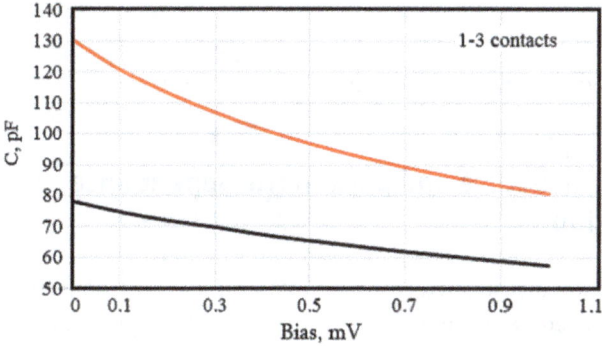

Figure 6.13: The farad-voltage characteristic of an individual Schottky barrier.

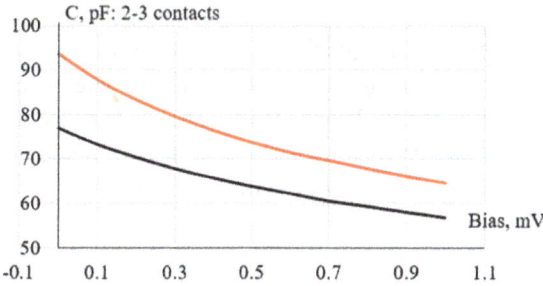

Figure 6.14: The farad-voltage characteristic of an individual potential barrier.

The dependence of the capacitance of the p^+-n-p^+ structure on the voltage of about 200 mV (Figure 6.15) has a maximum when the surface barrier is reverse-biased and the rear barrier is forward-biased both in the dark and under light exposure.

The mentioned voltage is spent on eliminating the difference between the oppositely directed potential barriers. Then, the further increase in the voltage widens the near-surface barrier and reduces the capacitance.

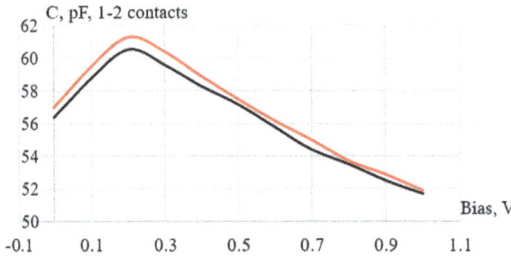

Figure 6.15: The farad-voltage characteristic of the p⁺-n-p⁺ structure.

6.4 p⁺(PtSi)-n(Si)-p⁺(Si) structure and parametric dependence of absorbed radiation

The relevant expressions for the investigated structure establish that the connection between the absorption coefficient a of the p⁺-n-p⁺ silicon structure (Figure 6.5), the photocurrent I_f and the intensity F of the quanta incident on the photosensitive surface, the applied external bias voltage V, and the position of the minimum x_m of the potential energy of the electrons in the conduction band of the structure were obtained in the following manner.

The structure in Figure 6.5 has a rear barrier region with the p⁺ conduction. As a result, the diffusion current of the electrons may flow to the base and become a part of the public current. The diffusion current can be determined by defining the n_p density of the holes, that is, of the minority charge carriers in the p-type semiconductor, by solving the one-dimensional diffusion equation stated below (S. Kh. Khudaverdyan et al., 2005; Sze et al., 2021):

$$\frac{\partial^2 n_p}{\partial x^2} - \frac{n_p}{L_n^2} = \frac{n_0}{L_n^2} - \frac{G(x)}{D_n} \tag{6.2}$$

and

$$L_n = \sqrt{D_n \tau_n}$$

where $G(x) = F_0 a e^{-ax}$ is the generation rate of the electron-hole pairs, F_0 is the total flux of the photons incident on the photosensitive surface, a is the absorption coefficient of the electromagnetic radiation, L_n is the diffusion length of the electrons in the n-base, D_n is the diffusion coefficient, τ_n is the lifetime of the nonequilibrium electrons, and n_{p0} is the density of the equilibrium electrons. Equation (6.1) is solved under the following boundary conditions: $n_p = n_{p0}$ when $x = \infty$, and $n_p = 0$ when $x = d$. First, the corresponding homogeneous equation should be considered:

$$\frac{\partial^2 n_p}{\partial x^2} - \frac{n_p}{L_n^2} = 0 \tag{6.3}$$

for which the characteristic equation is $k^2 - (1/L_n^2) = 0$, which has the roots $k_1 = 1/L_n$ and $k_2 = -(1/L_n)$.

$k_1 \neq k_2$; thus, the general solution of the homogeneous equation can be presented as $n_{p1} = C_1 e^{-x/L_n} + C_1 e^{x/L_n}$. Here, $C_2 = 0$ since, when $x = \infty$, then $n_{p1} = \infty$, which has no physical meaning. Thus,

$$n_{p1} = C_1 e^{-x/L_n} \tag{6.4}$$

The particular solution for eq. (6.2) should be found in the form of the product e^{-ax} and the polynomial with unknown coefficients

$$n_{p2} = A e^{-ax} + B \tag{6.5}$$

where the coefficients A and B of the equation are determined by the method of uncertainty coefficients. For that, eq. (6.5) is differentiated and, as a result, the following is obtained:

$$\frac{\partial n_p}{\partial x} = -A a e^{-ax}, \quad \frac{\partial^2 n_p}{\partial x^2} = A a^2 e^{-ax} \tag{6.6}$$

By inserting expressions (6.5) and (6.6) into eq. (6.2), the result will be

$$A a^2 e^{-ax} - \frac{A}{L_n^2} e^{-ax} - \frac{B}{L_n^2} = \frac{\alpha F_0}{D_n} e^{-ax} - \frac{n_{p0}}{L_n^2} \tag{6.7}$$

and therefore

$$\frac{B}{L_n^2} = \frac{n_{p0}}{L_n^2}, \quad A\left(\frac{a^2 L_n^2 - 1}{L_n^2}\right) = \frac{\alpha F_0}{D_n}, \quad A = \left(\frac{F_0}{D_n}\right)\frac{\alpha L_n^2}{1 - a^2 L_n^2} \tag{6.8}$$

and eq. (6.5) will take the form that is the general solution of eq. (6.2):

$$n_p = C_1 e^{-x/L_n} + n_{p0} + \left(\frac{F_0}{D_n}\right)\frac{\alpha L_n^2}{1 - a^2 L_n^2} e^{-ax} \tag{6.9}$$

C_1 is determined from the boundary conditions at $n_p = 0$ when $x = d$

$$C_1 e^{-d/L_n} = -n_{p0} - \left(\frac{F_0}{D_n}\right)\frac{\alpha L_n^2}{1 - a^2 L_n^2} e^{-ad} \tag{6.10}$$

Hence,

$$C_1 = -\left(n_{p0} + \left(\frac{F_0}{D_n}\right)\frac{\alpha L_n^2}{1-\alpha^2 L_n^2}e^{-ad}\right)e^{d/L_n} \tag{6.11}$$

As a result, the following equations for the density of the minority charge carriers in the n-region of the semiconductor and the corresponding diffusion current will be obtained:

$$n_p = n_{p0} - \left(n_{p0} + \left(\frac{F_0}{D_n}\right)\frac{\alpha L_n^2}{1-\alpha^2 L_n^2}e^{-ad}\right)e^{d-x/L_n} + \left(\frac{F_0}{D_n}\right)\frac{\alpha L_n^2}{1-\alpha^2 L_n^2}e^{-ax} \tag{6.12}$$

$$j_{ndif} = qD_n\frac{dn_p}{dx} = \left(\frac{qD_n n_{p0}}{L_n} + \frac{q\alpha L_n F_0}{1-\alpha^2 L_n^2}e^{-ad}\right)e^{d-x/L_n} + \frac{qF_0\alpha^2 L_n^2}{1-\alpha^2 L_n^2}e^{-ax} \tag{6.13}$$

At the point $x = d$, the expression for the photocurrent diffusion component is

$$j_{ndif} = qn_{p0}\frac{D_n}{L_n} + qF_0\frac{\alpha L_n}{1+\alpha L_n}e^{-ad} \tag{6.14}$$

The quanta absorbed between the surface and the point x_m provide the photocurrent of the surface barrier. For an individual wave, its absorption coefficient and the applied external voltage, it has the following form (S. Kh. Khudaverdyan et al., 2005):

$$I_1 = qSF_0\left(1 - e^{-\alpha_i x_{mj}}\right) \tag{6.15}$$

and the photocurrent generated by the rear barrier is

$$I_2 = qSF_0\left(e^{-\alpha_i x_{mj}} - e^{-\alpha_i d}\right) \tag{6.16}$$

The public photocurrent conditioned by a single wave is the difference between those two photocurrents and has the following form:

$$I = qSF_0\left(1 - 2e^{-\alpha_i x_{mj}} - e^{-\alpha_i d}\right) \tag{6.17}$$

The diffusion current in the rear p^* region is determined by expression (6.14).

Under normal working conditions, the term containing n_{p0} can be neglected since it is considerably smaller than the second term.

Thus, the diffusion component of the term containing n_{p0} will be added to the expression of the photocurrent, and the photocurrent generated by one wave will have the form

$$I_{Phij} = qSF_0\left(1 - 2e^{-\alpha_i x_{mj}} - \frac{e^{-\alpha_i d}}{1+\alpha L_n}\right) \tag{6.18}$$

and the total photocurrent generated by all the waves of the integral radiation flux will have the form

$$\sum_{ij} I_{Phij} = qS \sum_{ij} F_{0i}\left(1 - 2e^{-a_i x_{mj}} - \frac{e^{-a_i d}}{1 + aL_n}\right) \tag{6.19}$$

where $(i = 1, 2, 3, \ldots)$ changes with the change of the radiation wavelength in the integral flux, $(j = 1, 2, 3, \ldots)$ changes with the change of the bias voltage, $F_0(\lambda_i)$ is the total flux of the photons of the incident wave with the length λ_i, and w is the range width of the generation of the diffusion current.

If the depth x_m corresponds to the most deeply penetrated wave from the integral flux, then a small voltage change will correspond to such a change in x_m within which only that wave will be absorbed. Having the experimental values of the photocurrent changing according to x_m, it will be possible to determine the absorption coefficient of the wave a. Since the most deeply penetrated wave reaches x_m, the three values of the photocurrent I_1, I_2, I_3 corresponding to the three neighboring values of the voltage with the difference of 1 mV will be conditioned only by the absorption of that wave. Thus, taking into account the above, in the absence of the diffusion component of the photocurrent, the transcendent equation for a will be obtained from expression (6.17):

$$\frac{I_1 - I_2}{I_2 - I_3} = \frac{e^{-a x_{m2}} - e^{-a x_{m1}}}{e^{-a x_{m3}} - e^{-a x_{m2}}} \tag{6.20}$$

By denoting $A = (I_1 - I_2)/(I_2 - I_3)$, we will have $A = (e^{-a x_{m2}} - e^{-a x_{m1}})/(e^{-a x_{m3}} - e^{-a x_{m2}})$.

Since the experimental values of the current are of the order of μA, it makes sense to carry out the transformation of the Maclaurin series around the point "0." Thus, for x_{m1}, x_{m2}, x_{m3} it will be

$$e^{-a x_{m1}} = 1 - a x_1 + \frac{a^2 x_{m1}^2}{2}, \quad e^{-a x_{m2}} = 1 - a x_2 + \frac{a^2 x_{m2}^2}{2}, \quad e^{-a x_{m2}} = 1 - a x_2 + \frac{a^2 x_{m2}^2}{2} \tag{6.21}$$

With the help of those transformations, we will receive

$$\frac{e^{-a x_{m2}} - e^{-a x_{m1}}}{e^{-a x_{m3}} - e^{-a x_{m2}}} = A = \frac{x_{m1} - x_{m2} - a\left(\frac{x_{m1}^2 - x_{m2}^2}{2}\right)}{x_{m2} - x_{m3} - a\left(\frac{x_{m2}^2 - x_{m3}^2}{2}\right)} \tag{6.22}$$

Thus, the finite equation will be replaced by the following approximation:

$$A(x_{m2} - x_{m3}) - x_{m1} + x_{m2} = a\left[\frac{A\left(x_{m2}^2 - x_{m3}^2\right) - x_{m1}^2 + x_{m2}^2}{2}\right] \tag{6.23}$$

and the equation determining a will be obtained:

$$a = \frac{2A(x_{m2} - x_{m3}) - x_{m1} + x_{m2}}{A\left(x_{m2}^2 - x_{m3}^2\right) - x_{m1}^2 + x_{m2}^2} \tag{6.24}$$

The diffusion component in the expression for the current is shrinking in the expression for A, thus leading to expression (6.24) for determining the absorption coefficient. Therefore, the highly approximate equation previously used for determining the absorption coefficient (S. Khudaverdyan, Avetsiyan, et al., 2013; S. Khudaverdyan, Khachatryan, et al., 2013a) will be

$$a_i = \frac{1}{\Delta x} \ln \frac{I_2}{I_1} \tag{6.25}$$

where $\Delta x = x_{m1} - x_{m2}$ can be replaced by expression (6.24) derived from the cause-and-effect relationships of the photodetector structure. Then, the expression for determining α will be used to ensure a more accurate determination of the spectral dependence. Having α and the experimental data of the photocurrent, it will be possible to determine the intensity of the given wave incident on the photosensitive surface of the photodetector with the help of the following expression:

$$F_{0i} = \frac{I_i}{\left(1 - 2e^{-a_i x_{mj}}\right) + \frac{e^{-a_i d}}{1 + a_i w}} \tag{6.26}$$

It will make it possible to build the dependence of the photocurrent generated by a given wave on x_m, subtracting it from the total photocurrent experimentally obtained

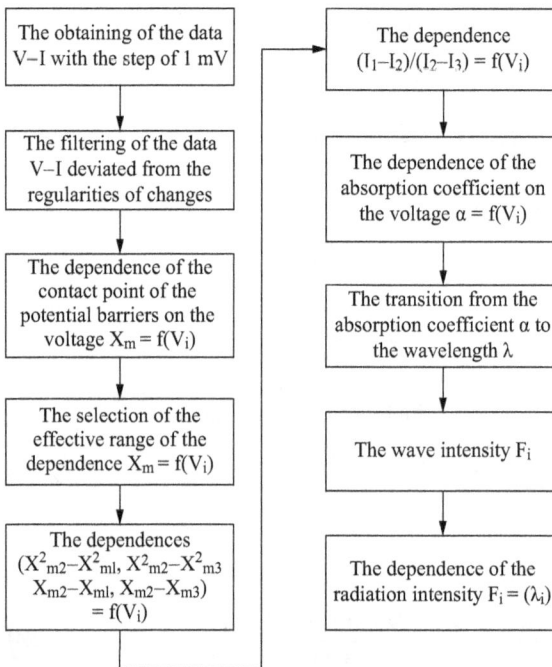

Figure 6.16: The simplified diagram of the algorithm of the spectral analysis.

for every voltage, and a series of new photocurrents without the photocurrent created by a determined wave. Now, a new small voltage change will move x_m toward the surface, and the next wave will get involved in the registration region depending on the penetration depth. It will make it possible to determine the length and the intensity of the next wave. The step-by-step repetition of the cycle will help to obtain the spectral dependence of the radiation intensity. The experimental research was carried out by employing different radiation sources. Figure 6.16 shows the simplified diagram of the algorithm of the spectral analysis.

6.5 Summary

– The wave absorption helped to determine the regularities of the current change using the different waves of the same intensity. All the curves have the change of the current sign corresponding to the difference of the potential barriers of 0.19 mV.
– The saturation photocurrent created by the shortwave (blue light) in the case of the near-surface barrier is greater than that created by the longwave (green and red lights), which is due to the redistribution of the intensities of the waves with the corresponding absorption depths between the barriers.
– Very small dark currents (up to the order of 10 pA) are obtained. It makes it possible to register very low intensities ($\sim 10^8$ sq/cm^2 s) or powers ($\sim 3 \times 10^{-11}$ W) and carry out the spectral analysis of weak optical signals.
– The algorithm for spectral-selective sensitivity includes the expression for the exact value of the absorption coefficient obtained via solving the transcendental equation. This ensures the high resolution of the spectral-selective sensitivity (up to 1 nm).

Part 5: **Research results**

Chapter 7
Performance assessment of photodetectors

7.1 Introduction

Performance assessment of photodetectors involves exhaustively evaluating various parameters that determine their efficiency and suitability for specific applications. Some of the key parameters commonly considered are responsivity and spectral response, dark current, noise equivalent power (NEP), bandwidth, linearity, dynamic range, rise/fall time, and temperature dependence. The performance assessment often involves a combination of theoretical analysis, laboratory experiments, and real-world testing to evaluate these parameters and determine the photodetector's suitability for specific applications. Additionally, factors like stability, reliability, and long-term performance are all important considerations, especially for applications requiring continuous operation over extended periods. Some of the most relevant assessment parameters are described here, while the previous chapters have provided a theoretical basis for such parameters.

7.2 Performance capabilities of the spectral-selective sensitivity of the photodetectors

Table 7.1 shows the spectral parameters of the calibrated beam of the xenon lamp used during the experimentation of the photodetectors for this investigation. The measurements were carried out by the measuring device B2912A. The photodiode S120A with a spectral range of 400–1,000 nm was used during the calibration. It is for this reason that two values smaller than 400 nm may be accepted with a relatively large error fraction for 300 and 350 nm. In qualitative assessments, this can be taken as noncritical. The diffraction grating used during the spectral measurements provided the spectral accuracy of $\Delta\lambda = 10$.

Figure 7.1 shows the spectral distribution of the beam intensity of the xenon lamp obtained with the help of the algorithm described above and the data of the experimental current-voltage characteristics of the structure presented in Figure 5.5. It roughly corresponds to Figure 7.2 built with the help of the data presented in Table 7.1. The maxima in Figure 7.1, missing in Figure 7.2, are the result of the minimum step change in the wavelength and are closer to the technical data of the xenon lamp (E4U, 2023).

Further noise reduction and the structure upgrade will lead to an increase in accuracy. The studies were also carried out with the help of radiation sources, such as LDs-813SRC-J14 (AlGaInP), L53GC (GaP), and LL-304B-B4-GD (InGaN). The numerical data of the current-voltage characteristics of the photodetector were the input data of

https://doi.org/10.1515/9783111428024-008

Table 7.1: Spectral parameters of the calibrated beam of the xenon lamp.

Wavelength (Nm)	hv (eV)	Power, µW	Number of quanta on the surface: F_0 (sq/cm² S)	Absorption coefficient: α (cm⁻¹)
300	4.14	11.84	1.787E + 13	1.73×10^6
350	3.54	59.6	1.052E + 14	1.04×10^6
400	3.1	274.4	5.532E + 14	9.52×10^4
450	2.7	382	8.843E + 14	2.55×10^4
500	2.5	391.4	9.785E + 14	1.11×10^4
550	2.26	371	1.026E + 15	6.39×10^3
600	2.06	359.2	1.09E + 15	4.14×10^3
650	1.91	320.3	1.048E + 15	2.81×10^3
700	1.77	245	8.651E + 14	1.90×10^3
750	1.65	216.2	8.189E + 14	1.30×10^3
800	1.55	186.3	7.512E + 14	8.50×10^2
850	1.55	123.9	4.996E + 14	5.35×10^2
900	1.4	847.6	3.784E + 15	3.06×10^2
950	1.3	841.9	4.048E + 15	1.57×10^2
1,000	1.24	870.9	4.39E + 15	6.40×10^1

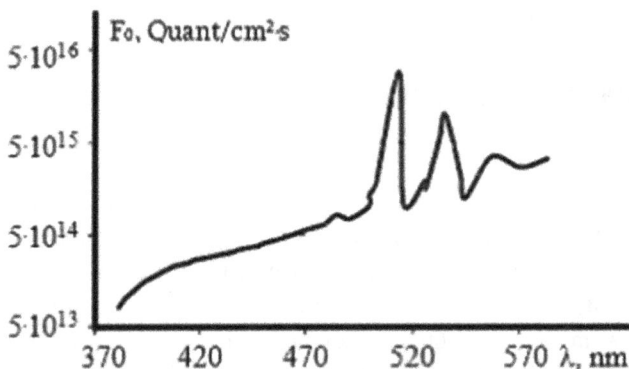

Figure 7.1: The spectral distribution of the beam intensity of the xenon lamp.

the algorithm. The automatic mode of the algorithm implementation in the Excel environment was provided. The output data in the form of spectral distribution of LED intensities derived from the experiments are presented in Figures 7.3–7.5.

The maxima in them have the wavelength values $\lambda_{max} = 650$ nm, $\lambda_{max} = 565$ nm, and $\lambda_{max} = 462$ nm, respectively, which are very close to the technical data mentioned above. Comparing Figures 7.3–7.5, the following regularities are observed. The shorter the wavelength and the lower the intensity of the beam, the greater the size of the "fluctuations" of the spectral curve (the oscillations in Figure 7.3 are smaller than those in Figures 7.4 and 7.5). It can be assumed that it is the result of the small photocurrent values that are close to the noises.

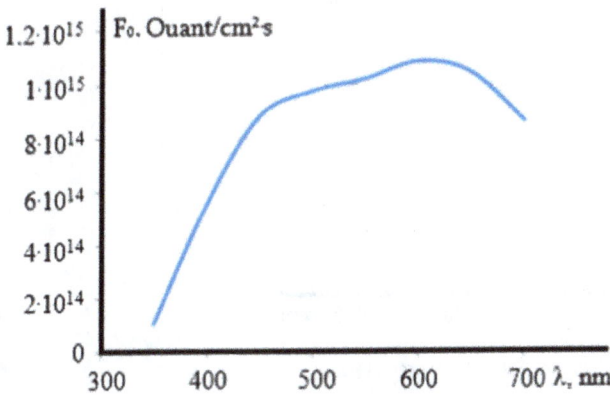

Figure 7.2: The spectral distribution of the beam intensity of the xenon lamp measured by the Thorlabs PM100 measuring device.

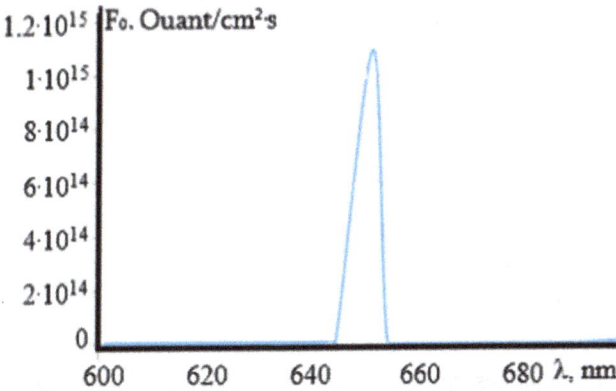

Figure 7.3: The spectrum obtained by the LDs-813SRC-J14 (AlGaInP) red LED.

The spectral curves obtained with the relative units given in Figure 7.6 have a narrow spectrum in the regions of blue and red rays compared to the reference ones. In the case of the absorption of green rays, the spectrum is close to the reference one.

It can be explained by the absorption depth of the corresponding rays. Figure 5.13 shows that the blue ray ($\lambda_{max} = 462$ nm) is completely absorbed (the residual intensity is close to zero), the red ray ($\lambda_{max} = 660$ nm) has a large unabsorbed residual intensity, and the green ray ($\lambda_{max} = 565$ nm) has an intermediate position, that is, it is partially absorbed and partially passed through.

Therefore, the green beam is more actively involved in the modulation of the widths of the potential barriers with the help of the external bias voltage. Figure 7.7 shows the relative spectral dependence of the intensity of the white LED LL-1003WC2D-W2-1B, and

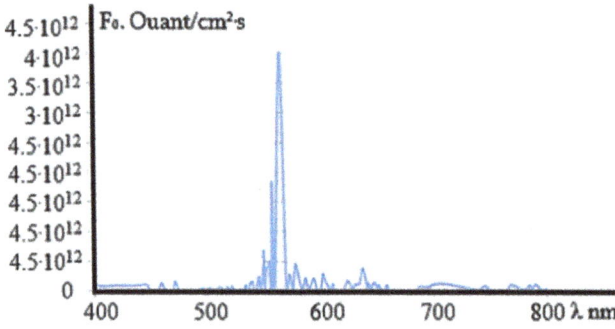

Figure 7.4: The spectrum obtained by the L53GC (GaP) green LED.

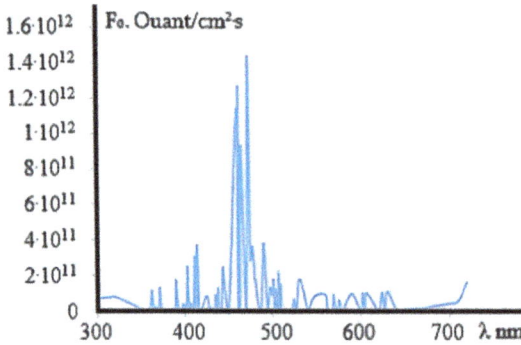

Figure 7.5: The spectrum obtained by the LL-304B-B4-GD (InGaN) blue LED.

Figure 7.8 presents the experimental absorption spectrum of its intensity. It roughly repeats the reference dependence shown in Figure 7.8.

The compared spectra are close to each other. The deflection is 10–30 nm. It will allow the consideration of the possibility of spectral analysis with the help of the developed photodetectors and the prospects of the development of spectrophotometers based on them. The selective sensitivity of the investigated structures covers the spectrum ranging from ultraviolet (UV) to near-infrared (IR). As is known, this part of the spectrum is used to detect hazardous substances in the environment. Therefore, the investigated photodetectors, with the help of the appropriate algorithm, claim to replace the well-known spectrophotometers and gain an advantage in price, reliability, and size.

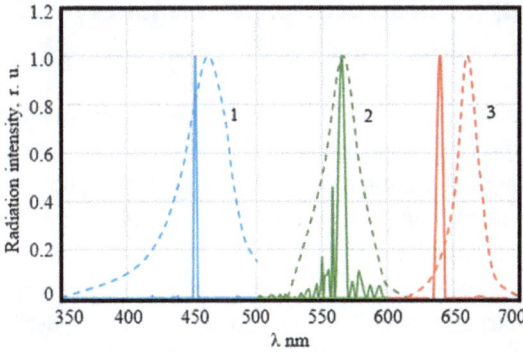

Figure 7.6: The reference (dotted lines) and experimental (solid lines) spectral dependences of (1) LL-304B-B4-GD (InGaN), (2) L53GC (GaP), and (3) LDs-813SRC-J14 (AlGaInP) red LEDs.

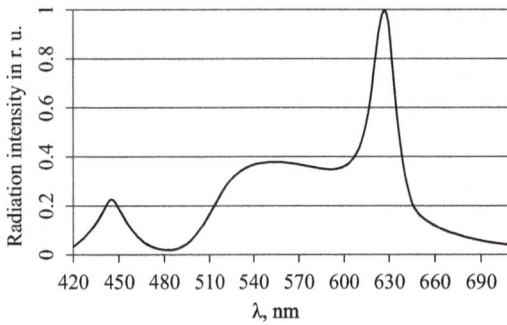

Figure 7.7: The relative spectral dependence of the intensity of the white LED LL-1003WC2D-W2-1B.

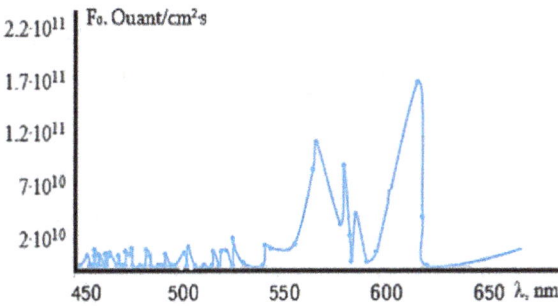

Figure 7.8: The absorption spectrum of the intensity of the white LED LL-1003WC2D-W2-1B.

7.3 Noises in the $p^+(PtSi)$-$n(Si)$-$p^+(Si)$ structure: performance and threshold photosensitivity

The functional capabilities of the photodetectors that meet modern requirements depend on their noise resistance and speed. The first determines the threshold photosensitivity, and therefore, the ability of the photodetector to register the smallest informative signals and the smallest wave deflections in the spectrometric solutions, and the second determines the ability to follow the fast processes.

Generally, in the semiconductor photodetectors, the noises that are due to the density fluctuations of the current charge carriers lead to the generation of the oscillating electromotive force of the noise. The processes like the drift of the charge carriers in the electric field and the diffusion in the structure base, the chaotic thermal motion, the trapping, the recombination, the scattering on the acoustic and optical phonons, deficiencies and charge carriers, the obstacle surmounting, and so on are the main causes of noise (Sze et al., 2021; I. Vikulin & Stafeev, 1990).

At the output of the photodetector, the absorbed optical and background radiations create a current consisting of dark, light, and background signals. Together with the thermal noises, they create an output signal and noise by the matching method. The mentioned summed signal at the output is the consequence of the splitting and the drifting of the electron-hole pairs generated by the absorption of the optical signal and the background beam by the electric fields of the barriers toward the opposite sides of the barriers. The output current includes the signal, dark, background, and noise currents. The background current can be excluded from the above-mentioned currents by creating the necessary measurement conditions. The noise currents remain insurmountable to some extent.

The generation and the recombination noises or the shot noises, as well as the thermal and frequency noises, are the main noises of the semiconductor photodetectors (Sze et al., 2021). The photodetector noises are divided into low-frequency (excess), white, thermal, and shot noises. The excess noises play almost no role in the fast Schottky diodes. At frequencies higher than $f \sim 100$–$1,000$ Hz, they strongly decrease and are screened by the generation-recombination and thermal noises. The thermal noises have a low level. Thus, excess noises are the main noises in the semiconductor photodetectors. They do not depend on the low and medium frequencies and have a "white" spectrum.

The mean square value of the current fluctuations of the shot noises is determined by the expression $\overline{I_{sh}^2} = 2qI\Delta f$ (Sze et al., 2021; I. Vikulin & Stafeev, 1990), where Δf is the frequency transmission passband and $I_{sum} = I_D + I_L$ is the sum of the dark and light currents. The threshold sensitivity of the photoreceiver is the minimum power of the radiation that will be detected against the noise background. For the monochromatic beam, it will be defined as (I. Vikulin & Stafeev, 1990)

$$P_{threshold} = (2q\Delta f I_{sum})^{1/2} / S_{i\lambda} \qquad (7.1)$$

If the photocurrent conditioned by the background radiation and the dark current conditioned by the thermal generation of the electron-hole pairs in the depleted layer are ignored, then it can be stated that the current flowing through the structure is the sum of the dark and light photocurrents. Under these conditions, the threshold sensitivity is the smallest detectable signal against the noise background and is determined by the signal-to-noise ratio equal to 1:

$$\frac{I_{sum}}{\sqrt{I_{sh}^2}} = \frac{P_{threshold} S_{i\lambda}}{I_{sh}^2} = 1 \qquad (7.2)$$

Where $S_{i\lambda} = I_L/P_L$ is the current photosensitivity of the structure, and $P_L = h\nu \times S \times F_0$ is the beam power?

The spectral sensitivity $S_{i\lambda}$ characterizes the minimum power of the detection signal. It is confined by the presence of noises. In the investigated structure, the current defined by eq. (6.17) is the current of the reverse-biased potential barrier at any polarity. It has a diffusion component and is obtained with the help of experimental measurements. By inserting it into eq. (7.1) and taking the photosensitivity ~0.5 A/W of the popular silicon photodiodes, the threshold sensitivity may be determined. The latter may also be subjected to the spectral analysis if eq. (6.18) is inserted into (7.1), where the diffusion current is neglected, since, according to Figure 5.13, the residual intensities of the wavelengths up to 550 nm are significantly small, are mainly absorbed in the base, and do not create the diffusion current. As a small quantity, the dark current may also be neglected (it does not exceed 10^{-10} A in the investigated samples). In that case, in the absence of the beam reflection, the expression for the change in the threshold sensitivity will be the following:

$$P_{threshold} = \left[2 S F_0 q^2 \Delta f \left(1 - 2e^{-a_i x_{mj}} \right) \right]^{1/2} / S_{i\lambda} \qquad (7.3)$$

In the investigated samples, $S = 0.02$ cm² and the spectral dependence of the threshold photosensitivity $P_{threshold}(a_i)$ hardly changes for different values of x_m in the unit layer of the frequency within the wavelength range of 250–500 nm, the threshold sensitivity range of ~7.8×10^{-14} W × Hz$^{1/2}$, and the voltage range of −1 to +1 V (active modulation range of x_m) when the beam intensity is ~1.5×10^{12} sq/cm² s. If the compensation of the oppositely directed barriers is considered, then that is a high coefficient.

In the expression of the current of the threshold sensitivity, at the absorption of the long waves, the diffusion component of the current should also be taken into account, since the residual intensities are quite large (see Figure 5.13). Thus, it is necessary to insert the expression for the current (6.18) into (7.1), and as a result, the threshold sensitivity will be

$$P_{\text{threshold}} = \left[2SF_0q^2\Delta f\left(1 - 2e^{-a_ix_{mj}} - \frac{e^{-a_id}}{1+aL_n}\right)\right]^{1/2}/S_{i\lambda} \qquad (7.4)$$

By inserting the above-mentioned parameter values in expression (7.3), the values for the threshold sensitivity will be $3.4\times10^{-14}\,W\times Hz^{1/2}$. This coefficient is about two times higher (smaller) than the coefficient without the diffusion component. Thus, the investigated structures have low noises and high threshold sensitivity.

7.4 Speed assessment

The inertia of the semiconductor photodetector is mainly conditioned by the time of diffusion (τ_{dif}) and drift (τ_{dr}) of the unbalanced charge carriers through the base, the flight time of the charge carriers through the p-n junction, and the time constant (Sze et al., 2021; I. Vikulin & Stafeev, 1990). In the investigated structure, the base is occupied by the space charges. The electrons drift through the base and have speed. In the test samples, the base is 2-μm wide. Therefore, the drift time will be $\tau_{\text{dr}} = (d/v_{\text{dr}}) \approx 2\times10^{-10}$ s. The flight time of the charge carriers in the silicon base is determined by their maximum speed $\tau_{fl} = (d/v_{\text{max}})\sim4\times10^{-11}$ s.

At the reverse bias voltage, the time constant of the photodetector $\tau_{RC} = RC$ is determined by the base resistance and the charge capacitance of the $p-n$ junction. In the calculations, the output resistance of the structure base is $R = d/Sqn\mu$ (Shalimova, 1985), where μ is the electron mobility and is equal to 1,200 cm^2/Vs in the silicon. By inserting the numerical values, $R = 0.058$ ohm is received.

The numerical values of the charge capacitance are presented in Figure 6.15, and by taking 60 pf, it will make $\sim3.5\times10^{-12}$ s for the time constant. Thus, the performance of the investigated structure is of the order of 10^{-10} s and is determined by the drift time. It is sufficient for solving the photospectrometric issues since the performance of modern spectrometers is 10^{-3} s (Green, 2008) (https://publiclab.org/wiki/fold able-spec accessed: 8 February 2024).

7.5 The efficiency assessment of the work in terms of the current demands

As mentioned above, to measure the corresponding intensities in the typical spectro-photometers, prisms, diffraction gratings, light filters, high-precision optical and mechanical devices, as well as a series of photodiodes or an *LC* matrix are used for the spatial dispersion of the integral beam waves. The main factors limiting their use are their size (more than several tens of centimeters) and their cost (in excess of ¢ =1,500 for the 1-nm spectral resolution).

The theoretical and practical research done in the direction of obtaining a new type of spectrophotometric photodetectors and presented in the monograph reveals the possibility of obtaining the beam spectrum via measuring the current and the voltage of the specially developed photodetector. A corresponding algorithm and a software package are developed for that purpose. The spectrometers based on such a photodetector (Figure 7.9) do not require any prisms or diffraction gratings. The spectrometers are expected to have a measurement spectral range of 200–1,000 nm and a spectral resolution of 1 nm. With such parameters, they are not inferior to the USB spectrophotometers. From this standpoint, the photodetector is a good option for creating spectrometers based on it, which will replace modern small-size spectrometers. It is important to note that due to the small size of the photodetector or the chip based on it, it can be integrated into the mobile phones giving them a new spectrometric function. The spectrometers of modern mobile phones are large (Figure 7.10).

The feasibility assessment of the Si technology. Si is the most widespread and economically available material in semiconductor manufacturing. However, the absorption capacity of Si covers only a small part of the spectrum, 200–1,000 nm.

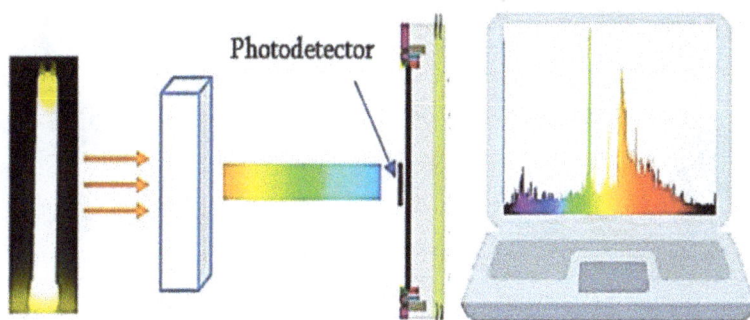

Photodetector

Figure 7.9: The diagram of the new spectrometer. A single semiconductor photodetector is used for obtaining the entire spectrum.

Further improvements can be made by studying the possibility of making detectors from other materials, which would extend the frequency range to the deep UV and mid-IR ranges. This technology will make it possible to make low-cost small-size spectrometers that can be integrated into smartphones.

The end result of the research is a device that can be used for solving environmental and health problems.

The main competitive advantages of the spectral analysis carried out by a single photodetector are:
-   The small size is achieved by removing the optical path and the diffraction grating.
-   No need for beam adjustment, as there are no optical elements requiring fine adjustment.

- High photosensitivity, as there are no light filters, optical paths, prisms, and diffraction gratings to reduce it.
- Low cost, as there are no optical elements, and the material consumption is low.
- The suggested spectrometers are in high demand due to the worldwide increasing demand for low-cost, fast, and reliable spectrophotometers. They are intended for the real-time monitoring of chemical and biological substances, food, and the environment. The wide range of spectrophotometer applications covers agriculture, veterinary medicine, nutrition, juice production, water supply management, environmental protection from hazardous substances and radiation, clinical diagnostics, drug monitoring, aerospace and mining industries, and so on.

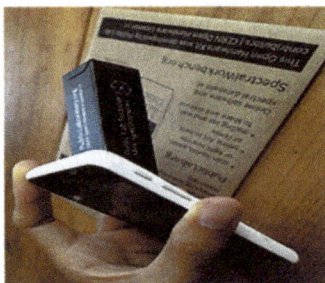

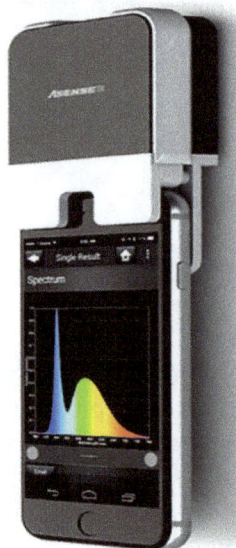

Figure 7.10: Examples of the spectrometers used in modern small-size smartphones. The left one is for professional use, and the right one is (Huo & Konstantatos, 2018) for amateur use.

7.6 Summary: a demand analysis of the proposed photodetectors

According to the research, the spectroscopy market is quite large. The development of a more efficient measurement device is a major demand factor and a necessity. The device can be improved under the laboratory conditions. It can be made smaller, more reliable, and with better parameters. The total spectroscopy market was about $ 7.5 billion in 2014; according to forecasts, it was approximately $ 0.550 billion in 2019 (https://www.researchandmarkets.com/reports/5157122/global-spectrophotome ter-market-2020-2024, accessed on 24 June 2022), and the Global Spectroscopy market reached $ 39 billion in 2022 (Atvars et al., 2019). It is estimated that the demand for spectroscopy in metallurgy, food, chemical, and automotive industries will continue to grow. A similar growth is expected in the fields of security and defense, and the scientific research is aimed at the discovery and application of new substances.

The spectroscopy market is quite competitive. There are many spectrometer manufacturers in the USA, Europe, and Japan. And yet, the production of basic elements such as light spectrometric detectors is under the control of Japanese companies (Hamamatsu, Sony, and Toshiba). The spectrometer manufacturing business will require significant investments from European companies. Currently, all the "know-how" and patents for manufacturing inexpensive detectors with excellent technical characteristics belong to Japanese companies. The only way to come into the market is to create a breakthrough technology.

When the monochromatic spectrometers were replaced by grid photodetectors and LCs, it gave a significant impetus to the spectroscopy market. The use of low-power processors, the absorbing properties of the material, and the low cost will make it possible to reach the production level. To date, in optical spectrometers, the spectral components are separated by the diffraction grating and are analyzed individually. The method is tested and works well. However, this research shows that the above-stated method can be improved by removing the optical path and the diffraction grating, and by replacing multiple photodiodes with a single photodetector. The core is a diode structure that, by performing mathematical calculations in real time by an inexpensive processor, can filter different radiation spectra depending on the absorption. As a result, the size of the spectrometer will be reduced, the costs will be decreased, and the device will become available to a larger section of its potential market.

Currently, the technological problem is solved by photodiode matrices (produced by Hamamatsu, Toshiba, Sony, etc.) and diffraction gratings (Horiba Jovin, Edmund Optics, Ibsen Photonics, etc.). In the near future, the Polytechnic University of Milan may come up with an alternative approach. However, the latter technology is aimed at the image processing market and does not provide the required resolution and measurement accuracy. Of course, many projects solve spectrometric problems, but none of them provide competitive technical characteristics. The most competitive market of spectrometers is the market of spectroradiometers. These devices are sold extensively. There are two main obstacles to successful commercialization: the market status quo and intellectual property protection. Therefore, it is necessary to prepare a prototype and show the advantages. The development of the intellectual property strategy is quite an important task from the perspective of the feasibility study.

Part 6: **On the semiconductor spectroscopy**

Chapter 8
Identification of emergent contaminants in transparent media

8.1 Introduction

Emergent contaminants in transparent media, such as water or air, pose significant challenges and concerns due to their potential impacts on human health, ecosystems, and the environment (Vaseashta et al., 2013, 2021, 2022; Vaseashta & Maftei, 2021). As scientific knowledge and analytical techniques advance, several researchers continue to identify new emergent contaminants and assess their potential risks. Several substances previously considered benign may be reevaluated based on emerging evidence of their environmental persistence or toxicity (S. Khudaverdyan et al., 2021; Vaseashta et al., 2013, 2021). Hence, it is essential to carry out a theoretical study of photoelectronic processes using devices, under investigation such as silicon (Si) n^+-p-n^+ structures. The contribution of various mechanisms of photon absorption to the total photocurrent is calculated. We investigated several mechanisms, including the influence of tunneling on spectral characteristics and the selective spectral photosensitivity of the samples under investigation. The relationship between the energy parameters of the absorbed waves and the structural parameters is elucidated. We derived expressions for the photocurrent with and without external diffusion current in the n^+-p-n^+ structures, as well as for the absorption coefficient. In some samples, we observed electron injection through the direct-mesh n-p junction, leading to enhanced spectral photocurrent. Conversely, in samples without injection amplification, there was an inversion of the spectral photocurrent sign. The inversion point exhibited linearity with the offset voltage, offering a means to determine unknown wavelengths. Mutual compensating transitions in Si structures resulted in a shift of the maximum spectral photosensitivity from the intrinsic (~850 nm) to shorter wavelengths, around 590 and 530 nm. Our study suggests that by selecting appropriate structural parameters, it is possible to more accurately obtain different short-wavelength spectral maxima for specific applications, such as detecting new and emerging contaminants in aqueous media.

8.2 State-of-the-art semiconductor spectroscopy

The importance of applied science has grown significantly, driven by humanity's need to address new challenges and solve issues related to health, biological and ecological safety, climate change, water and food monitoring, as well as advancements in military power and the exploration of the universe and the Earth. Many of these challenges are tackled using primary data recorders. Among them, spectral analysis of optical signals,

https://doi.org/10.1515/9783111428024-009

employing semiconductor photodetectors as primary sensors, holds significant importance. Enhancing photodetector parameters through novel spectrophotometric functions is a pressing task.

Traditionally, spectral analysis relies on devices containing monochromators, diffraction gratings, prisms, or optical filters, which can be inefficient and costly. To overcome these drawbacks, there is an urgent need to develop semiconductor photodetectors with selective spectral sensitivity for conducting spectral analysis. By employing such photodetectors in spectrophotometry, the reliance on diffraction gratings, prisms, and high-precision mechanical systems can be eliminated. This advancement promises high resolution, reliability, fast spectrum registration, affordability, and compact size. So far, the work carried out in this direction has remained at the research level, since it requires special conditions and sophisticated technology (Gergel et al., 2006; Seymour et al., 2014b; Vanyushin et al., 2005; Wachowiak et al., 2013). The research is in high demand these days (Technavio, 2017, 2020) since the identification of the composition of the environment in the field is safety critical. Besides, the use of such photodetectors in up-to-date multipurpose monitoring systems is of significant importance (Jiang et al., 2009; Normatov et al., 2015). Hence, several researchers have developed and studied two-barrier structures (S. Khudaverdyan et al., 2009, 2021; S. Khudaverdyan, Khachatryan, et al., 2013a; S. K. Khudaverdyan, 2003; S. Kh. Khudaverdyan et al., 2005). The structures are Si-based, featuring a vertically placed Si junction and oppositely directed p-n or n-p junctions. A novel physical principle applied in the photodetector ensures selective spectral sensitivity. These structures, along with those featuring two oppositely placed p-n or n-p junctions, exhibit several new features. Below are the research findings.

8.2.1 Innovative method of spectroscopy

A novel method is employed utilizing two vertically placed junctions, where the photocurrents generated by longitudinal illumination partially or completely compensate for each other. External voltage alters the widths of the depleted regions of the junctions, leading to changes in the fraction of absorbed quanta of each wavelength, thus resulting in variations in the photocurrents, one affecting the other. This process is also influenced by the absorption depth of the wave. By considering these factors, expressions for the resultant photocurrent are derived based on the energy parameters of the radiation and the structural parameters, enabling the development of an algorithm for spectral intensity distribution.

8.3 Photoelectronic processes: selective spectral sensitivity

To comprehend the mechanisms of photoelectronic processes in longitudinally illuminated Si two-barrier structures, it is crucial to assess the contribution of various photon absorption mechanisms to the total photocurrent. Utilizing the Einstein relation, we can determine the diffusion coefficients of electrons and holes in Si: $D_n = $ ~59 cm^2/s, $D_p = $ ~16 cm^2/s, respectively, and the diffusion lengths are $L_n = $ ~6 × 10^{-3} cm, $L_p = $ ~4 × 10^{-3} cm at the diffusion time of 10^{-6} s. In real samples, the diffusion time may be 10 times as much, which increases diffusion lengths several times.

In the examined structures, the base region typically ranges from 2 to 6 μm in thickness. This region is enveloped by the space charges of both barriers, resembling thin n-p junctions due to their significantly smaller width compared to the diffusion length of electrons and holes (Shalimova, 1985). Consequently, charge carriers traverse the space charge layer without recombination, implying the absence of recombination within the n-p junction regions.

In regions with a high density of ionized impurities, around 10^{-18} cm, the Fermi level closely approaches either the conduction band bottom or the valence band top. However, in the investigated structures, the level of degeneracy necessary for interband absorption is absent. Instead, there may be instances where radiation absorption disturbs the transition of an electron from the valence band to the conduction band, resulting in the formation of an exciton – a bound system of an electron and a hole – with an energy lower than the energy gap width (Shalimova, 1985). However, the formation of a stable exciton system typically occurs only at sufficiently low temperatures.

In cases where the exciton binding energy approaches the energy of lattice thermal vibrations, around 25.8 meV at 300 K, the exciton may dissociate, leading to the disappearance of corresponding lines in the absorption spectrum. While exciton states are typically observable only in deeply frozen bulk semiconductor samples, they are well-expressed at room temperature in thin-film nanoscale semiconductor structures. In such structures, the properties of excitons, including their binding energy, can be controlled by altering the size of nanostructures, enabling the manipulation of excitons in reduced dimensional structures (Sze et al., 2021) and the development of devices based on excitonic processes.

However, the investigated structures are significantly thicker than nanoscale structures, resulting in practically negligible exciton absorption at room temperature. Instead, absorption primarily occurs through mechanisms involving free carriers, impurities, and lattice vibrations, leading to an infrared absorption spectrum (at low temperatures) with low absorption coefficients. Consequently, relatively thick samples are required to achieve reasonable absorption levels.

8.3.1 Tunneling issues

According to the literature data (Green & Keevers, 1995), the absorption coefficient's numerical dependence on wavelength in Si has been extensively verified experimentally. This dependence encompasses all forms of absorption, contributing predominantly to the long-wavelength portion of the spectrum beyond the absorption edge, excluding bandgap absorption. Although the contribution of bandgap absorption is minor, its presence may marginally extend the main spectral distribution toward longer wavelengths. Considering this, in interpreting our experimental results, we have utilized the referenced dependence of the absorption coefficient on wavelength for Si, widely employed in the development of Si-based solar cells.

Furthermore, we separately discuss the influence of tunneling on the photoelectronic processes observed in our samples. It is well-established that the electric field arises from the gradients of the energy bands. In this context, electrons have the ability to tunnel through the triangular barrier (Figure 8.1a). The barrier height, denoted as E_g, and the thickness, denoted as d, can be calculated as follows:

$$d = E_g/qE \tag{8.1}$$

where E is the electric field strength.

With the increase in E, the slope increases and d decreases, and the tunneling probability increases. Upon the absorption of a quantum with the energy hv, the barrier thickness decreases to the value d' (Figure 8.1.b):

$$d' = (E_g - hv/qE) \tag{8.2}$$

As a consequence, the likelihood of the tunneling junction increases. In the presence of a strong electric field, the reduction in barrier thickness is analogous to a decrease in the energy gap width. This results in a shift of the absorption edge toward lower energies, corresponding to longer wavelengths. It is established that electron tunneling through the potential barrier in Si initiates at thicknesses less than $d \sim 8E^{-7}$ cm. When the voltage incident on the base $V = 1V$ and the base thickness equals 2μm, the field strength $E = 1/2E^{-4} = 5,000$ V/cm. Thus, from eq. (8.1), the barrier thickness $d' = (E_g - hv/qE)$, and $(E_g - hv) = E = 8E^{-7} \times 5,000 = 0.004$ eV. It is by this small amount that the energy gap width decreases. With the voltage incident on the base equal to $2V$, the decrease will make 0.008 eV, etc. Consequently, the absorption edge shifts toward lower energies due to tunneling at high voltages and narrow bases.

Indeed, while the shift resulting from tunneling is noticeable, it remains relatively small. Therefore, it is accurate to conclude that in the structures under investigation, the primary mechanism is predominantly bandgap absorption.

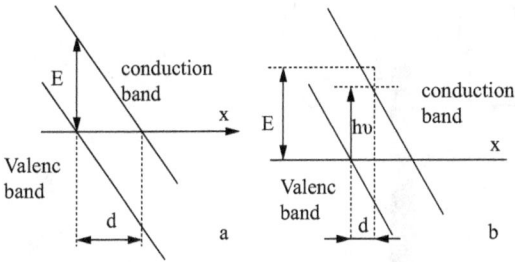

Figure 8.1: (a) The electrons can tunnel through the triangular barrier and (b) the absorption of a quantum with the energy hv decreases the barrier thickness to the value d'.

8.4 Interrelation of structural parameters

To solve the given problem, we need to derive an expression that relates the structural, energy, and technological parameters of the photodetector. Specifically, we examine p^+-n-p^+ and n^+-p-n^+ photodetectors, where the base is enveloped by $x_m - x_0$ and $d - x_m$ depleted layers of oppositely directed potential barriers, respectively (see Figure 8.2, where the n^+-p-n^+ structures denote the connection point of the barriers). Here, we also consider the width of the near-surface layer with a thickness x_0. The potential distribution is governed by the solution of the Poisson equation, which establishes the relationship between the field potential $V(x)$ and the volume density of charges N_a that generate this field (Figure 8.2).

The Poisson equation takes the following form:

$$\frac{d^2 V(x)}{dx^2} = -\frac{\rho}{\varepsilon \varepsilon_0} \tag{8.3}$$

We proceed from the potential $V(x)$ to the potential energy of electrons $\varphi(x)$, $\varphi(x) = -qV(x)$ in the Poisson equation. Since $\rho = qN_a$, we receive the following:

$$\frac{d^2 \varphi}{dx^2} = -\frac{q^2 N_a}{\varepsilon \varepsilon_0} \tag{8.4}$$

where N_a is the p-type impurity density, ε is the relative permeability of the substance, ε_0 is the permittivity of free space, and q is the electron charge.

The boundary conditions for this equation are $(|d\varphi/dx| = 0)$ at $x = x_m$ (x_m is the maximum of the potential energy of holes) and $\varphi(x) = \varphi_{b1}$ at $x = x_0$ (Figure 8.2). Taking into account the integrated equation (8.4) and that at $x = d$, when the external bias voltage is present, $\varphi(d) = \varphi_{b2} + qV$ (Figure 8.3), so we obtain an expression for x_m and $d - x_m$ depending on the external bias voltage:

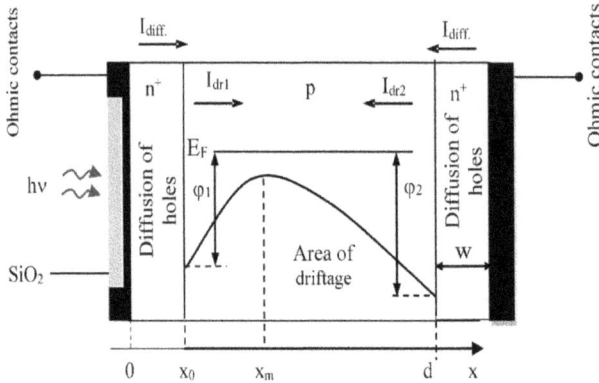

Figure 8.2: The n⁺-p-n⁺ structure. The distribution of the potential energy of holes in the valence band and the direction of the photocurrents (S. Khudaverdyan et al., 2022).

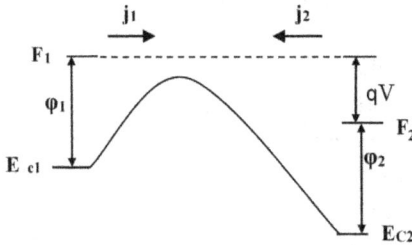

Figure 8.3: The energy change of the valence band under the influence of the external voltage.

$$x_m = \frac{d+x_0}{2} - \frac{\varepsilon\varepsilon_0(\Delta\varphi - qV)}{N_d q^2(d-x_0)} \ , \ d - x_m = \frac{d-x_0}{2} + \frac{\varepsilon\varepsilon_0(\Delta\varphi - qV)}{N_d q^2(d-x_0)} \tag{8.5}$$

If the near-surface n⁺ layer is thin, then $x_0 = 0$ and expression (8.5) will have the previously obtained form (S. Khudaverdyan et al., 2021):

$$d - x_m = \frac{d}{2} + \frac{\varepsilon\varepsilon_0(\Delta\varphi + q(-V))}{q^2 N_a d} \ \text{ or } \ x_m = \frac{d}{2} - \frac{\varepsilon\varepsilon_0(\Delta\varphi + q(-V))}{q^2 N_a d} \tag{8.6}$$

By eqs. (8.5) and (8.6), we can find the limiting values of the voltage at $x_m = d$ and $x = x_0$.

8.4.1 Optimal design of photodetectors

Taking into account the current-voltage characteristics (Figure 8.4), during the radiation absorption of the blue LED, an abrupt change in the photocurrent occurs within the voltage range of 0–0.24 V. This transition happens when the difference in potential barrier heights is overcome, resulting in the near-surface barrier becoming equal to the rear one.

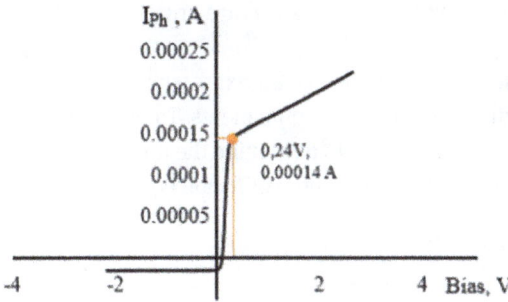

Figure 8.4: Light current-voltage characteristic at the radiation absorption of the blue LED.

As the near-surface barrier expands, the increase in photocurrent slows down, likely due to a decrease in the depth of absorbed quanta. Table 8.1 presents the absorption depths of the wavelengths corresponding to the peaks of blue, green, and red LEDs.

Table 8.1: Parameters of the waves of blue, green, and red radiations.

Wavelength (nm)	Absorption depth (nm)	Absorption coefficient (cm^{-1})
460	476	$2,1 \times 10^4$
560	1,730	5.78×10^3
660	3,870	5.58×10^3

It is evident that with a base width of $2\,\mu m$, the blue and green radiations can be efficiently absorbed, whereas for the red radiation, a base width of approximately $4\,\mu m$ is required.

8.5 Deriving the photocurrent expression for selected structures

Taking into account the exponential law governing the absorption of the specified wavelength, the expressions for the drift currents (Figure 8.2) formed in the base of the examined structures are as follows:

$$I_{dr1} = qF_0 S(1 - e^{-\alpha x_m}) \tag{8.7}$$

$$I_{dr2} = qF_0 S(e^{-\alpha x_m} - e^{-\alpha d}) \tag{8.8}$$

where α is the absorption coefficient of the electromagnetic radiation, S is the photosensitive area, and $F_0 = P_{opt}(1 - R)/S\,h\nu$ is the total flux of the incident photons per unit area. Here, P_{opt} is the radiation power, R is the reflection coefficient, h is Planck's constant, ν is the frequency of the electromagnetic radiation, and q is the electron charge. The presence of an out-of-the-base region in the structure creates a diffusion component in the public photocurrent.

Neglecting the thermal generation current, we derive an expression for the total photocurrent by considering both the drift and diffusion components. To determine the diffusion photocurrent in the structure, it is necessary to determine the minority carrier density p_n in the n-semiconductor using the one-dimensional diffusion equation:

$$\frac{\partial^2 p_n}{\partial x^2} - \frac{p_n}{L_p^2} = \frac{p_{n0}}{L_p^2} - \frac{G(x)}{D_p} \tag{8.9}$$

where $L_p = \sqrt{D_p \tau_p}$ is the hole diffusion length in the n-region, $G(x) = F_0 a e^{-a x_m}$ is the hole-electron generation rate, D_p is the hole diffusion constant in the n-region, τ_p is the lifetime of the excess carriers (holes), and p_{n0} is the equilibrium concentration of the holes in the n-region.

The solution of eq. (8,3) under the boundary conditions $p_n = p_{n0}$ at $x = \infty$ and $p_n = 0$ at $x = d$ (p_n is the equilibrium concentration of the minority charge carriers in the n$^+$- region of Figure 8.2) has the form

$$I_{\text{diff}} = S \left(q p_{n0} \frac{D_p}{L_p} + q F_0 \frac{a L_p}{1 + a L_p} e^{-ad} \right) \tag{8.10}$$

Considering (8.7), (8.8), and (8.10), the expression for the total current through the structure takes the form

$$I_{\text{tot}} = I_{\text{dr1}} - I_{\text{dr2}} - I_{\text{diff}} = S q F_0 \left(1 - 2 e^{-a x_m} + \frac{e^{-ad}}{1 + a L_p} \right) - S q p_{n0} \frac{D_p}{L_p} \tag{8.11}$$

Since, in the case of the normal operation in (8.11), the term containing p_{n0} is considerably smaller than the second term, it can be neglected and expression (8.11) will take the form

$$I_{\text{tot}} = S q F_0 \left(1 - 2 e^{-a x_m} + \frac{e^{-ad}}{1 + a L_p} \right) \tag{8.12}$$

When irradiated by the integral flux (e.g., of the Sun), the expression for the photocurrent can be represented as

$$\sum_{i,j} I_{\text{Phi},j} =$$

$$= \sum_{i,j} I_{\text{dr1},i,j} - \sum_{i,j} I_{\text{dr2},i,j} - \sum_{i,j} I_{\text{difi},j} = S q \sum_i \sum_j F(\lambda_i) \left(1 - 2 e^{-a_i x_{mj}} + \frac{e^{-a_i d}}{1 + a_i w} \right) \tag{8.13}$$

where $i = 1, 2, 3, \ldots$ changes in the integral flux with the change in the emission wavelength and $j = 1, 2, 3, \ldots$ changes with the change in the bias voltage, (λ_i) is the total flux of the incident photons with the wavelength λ_i If in eq. (8.12), the width w of the n region is taken to be less than L_p, then the value of L_p can be replaced by w, which is presented in (8.13).

The spectral characteristics can be analyzed using eq. (8.12). Here, the external bias voltage V is determined by the dependence of x_m on the voltage. From Figure 8.2, it is evident that in the investigated two-barrier structure, unlike the single p-n junction, the dependence of the depleted regions' width on the bias voltage exhibits a linear behavior. Consequently, x_m can be uniformly varied from x_0 up to d by adjusting the external voltage.

By utilizing the experimental values of the photocurrents, it becomes feasible to determine the absorption coefficient of the wave α as a function of the change in x_m. If the depth of x_m corresponds to the most deeply penetrated wave from the integral flux, then a small voltage change will lead to a change in x_m within which only that wave will be absorbed. Since the most deeply penetrated wave reaches x_{m1}, x_{m2}, x_{m3}, the three values of the photocurrent I_1, I_2, I_3, corresponding to three voltage values with a difference of 1 mV, are solely conditioned by the absorption of that wave. Thus, in the absence of the diffusion component of the photocurrent, employing the method provided in eq. (6.12), we derive the following equation for the absorption coefficient:

$$\alpha = \frac{2A(x_{m2} - x_{m3}) - x_{m1} + x_{m2}}{A(x_{m2}^2 - x_{m3}^2) - x_{m1}^2 + x_{m2}^2} \tag{8.14}$$

where $A = (I_1 - I_2)/(I_2 - I_3)$.

When calculated, the diffusion component of the photocurrent gets canceled, and we again obtain eq. (8.14). Using α and the experimental data of the photocurrent, we determine the intensity of the absorbed wave with the help of eq. (8.12):

$$F_{0i} = \frac{I_i}{\left(1 - 2e^{-\alpha_i x_{mj}}\right) + \frac{e^{-\alpha_i d}}{1 + \alpha_i w}} \tag{8.15}$$

The subsequent step involves obtaining the dependence of the photocurrent of a given wave on x_m and subtracting it from the total photocurrent. Subsequently, a small incremental voltage change will shift toward the surface, and the registration region will encompass the next wave based on its penetration depth. This procedure facilitates determining the length and intensity of that wave. By iteratively repeating this cycle step-by-step, the spectral dependence of the absorbed wave can be obtained.

8.6 Case studies of three typical samples

8.6.1 Silicide-n-p

Positive photocurrents originate from the reverse-biased surface barrier. In the wavelength range of 350–600 nm, the photocurrent of the reverse-biased surface barrier, down to zero voltage, surpasses significantly the photocurrent of the reverse-biased rear barrier (Figure 8.5a). The peak in the shortwave region is around 530 nm. At positive voltages, more shortwave quanta are absorbed near the surface barrier compared to the rear barrier. Upon changing the voltage polarity, the behavior of the spectral photocurrent (which is negative) remains consistent. However, due to the fewer quanta reaching the rear junction region, the photocurrent magnitude is lower across all voltage values (Figure 8.5a). In the wavelength range of 600–1,000 nm, the quanta penetrate deeper, and the photocurrent of the rear junction becomes comparable to that of the surface barrier (Figure 8.5b). The primary peak is located at a wavelength of 830 nm, which closely aligns with the bandgap absorption of Si.

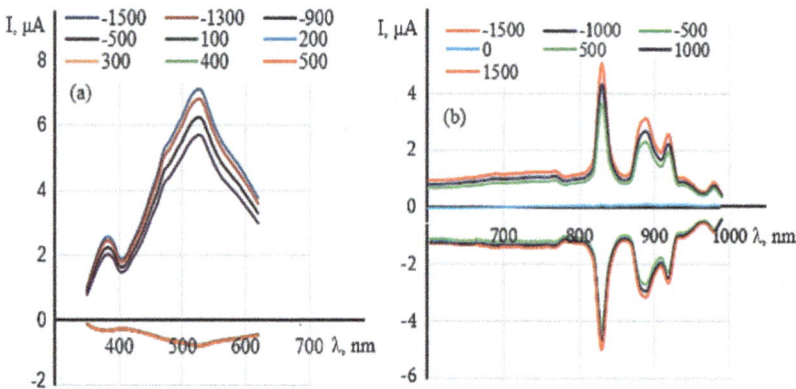

Figure 8.5: (a) Shortwave and (b) longwave current spectral photosensitivity for the silicide-p-n samples.

The spectral distribution of the current photosensitivity (as depicted in Figure 8.6a, b) reveals that the peak occurs in the vicinity of the bandgap absorption at 860 nm, with a value of 0.43 A/W. This sensitivity level is comparable to that of commercial Si photodiodes. At 530 nm, the photosensitivity measures 0.11 A/W.

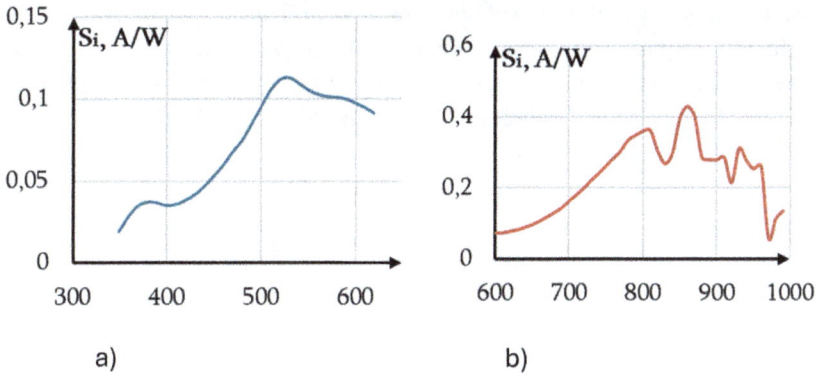

Figure 8.6: (a) Shortwave and **(b)** longwave current spectral photosensitivity for the silicide-p-n samples.

8.6.2 Silicon-n-p-n

In contrast to silicide-n-p structures, in this case, both in the shortwave and longwave regions, the negative photocurrents are comparable to the positive ones (Figure 8.7a, b). The peak in the longwave region occurs at a wavelength of 830 nm (Figure 8.7b), which is proximate to the bandgap absorption of Si. This observation is further supported by the spectrum of the current photosensitivity (Figure 8.8).

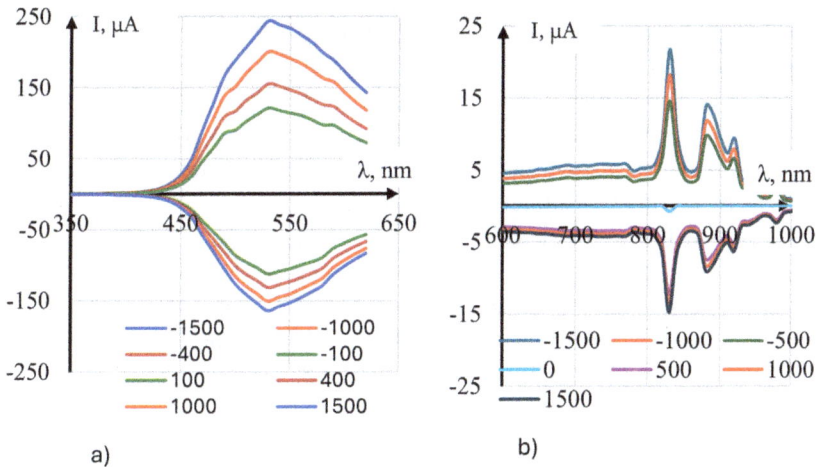

Figure 8.7: (a) Shortwave and **(b)** longwave spectral dependence of the photocurrent for the Si-n-p-n samples at different bias voltages V in mV.

The spectral distributions of the current photosensitivity (Figure 8.8a, b) reveal a maximum sensitivity of 4.1 A/W in the shortwave region at 560 nm, and 1.4 A/W in the longwave region at 830 nm. These values are notably high, particularly when compared to top-quality photodiode samples such as those from HAMAMATSU, which typically exhibit photosensitivity up to 0.7 A/W.

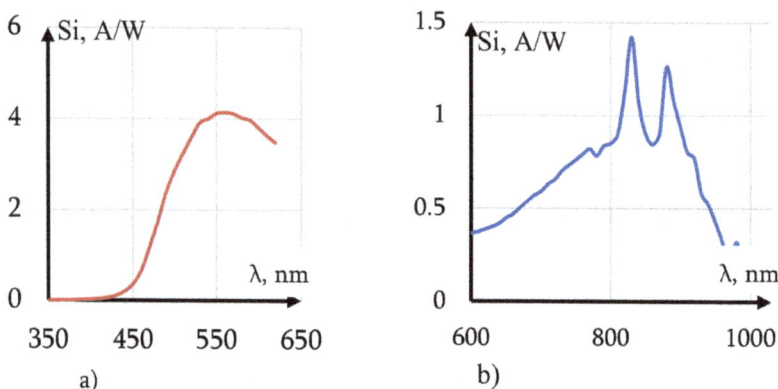

Figure 8.8: (a) Shortwave and **(b)** longwave current spectral photosensitivity for the Si-n-p-n samples.

8.6.3 Silicon1-n-p-n

The structure Si1-n-p-n exhibits distinct characteristics compared to previous cases. Unlike before, the negative photocurrent originates from the reverse-biased surface barrier (Figure 8.9). The change in the sign of the longwave photocurrent in the absence of bias indicates this (Figure 8.9b, yellow line). Shortwaves are absorbed near the surface barrier, generating a negative photocurrent, while longwaves are absorbed near the rear barrier, resulting in a positive photocurrent.

The spectral distributions of the current photosensitivity (Figure 8.10) demonstrate a maximum sensitivity of 0.03 A/W in the shortwave region at 480 nm, and 0.045 A/W in the longwave region at 830 nm. Consequently, there is a low photosensitivity, likely due to significant compensation of the photocurrents from both barriers. Hence, the Si-n-p-n and Si2-n-p-n structures exhibit abnormally high photosensitivities, possibly due to the presence of an internal amplification mechanism of the photocurrent. The speed of the Si-n-p-n and Si2-n-p-n photodetectors at a voltage of 20 mV is approximately 0.1–0.15 MHz during the pulse rise (Figure 8.11a), and about 0.04 MHz during the pulse drop (Figure 8.11b) of the photocurrent. This speed can increase with higher voltages.

The device can be recommended for the registration of the changes of not very transient optical signals.

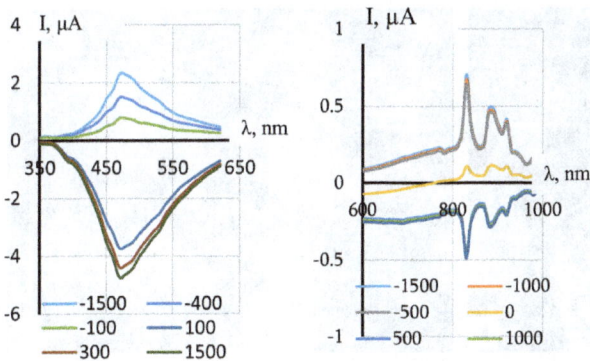

Figure 8.9: (a) Shortwave and **(b)** longwave current spectral photosensitivity for the Si1-n-p-n samples at different bias voltages V in mV.

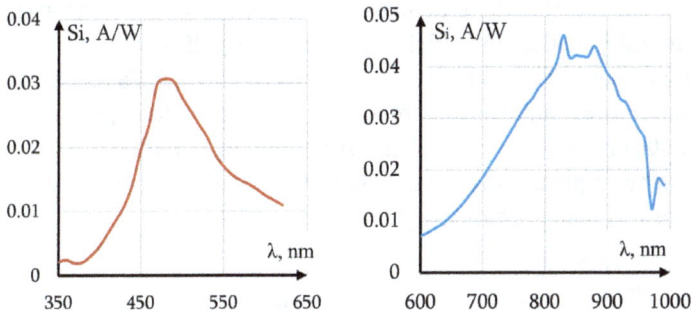

Figure 8.10: (a) Shortwave and **(b)** longwave current spectral photosensitivity for the Si1-n-p-n samples.

8.6.3.1 Photocurrent amplification mechanism

Initially, it was assumed that the radiation distribution in the samples follows the exponential law according to theory. This implies that by adjusting the voltage and the position of the junction point of the spectral photocurrent, it becomes possible to redistribute radiation absorptions between the barrier regions. It is essential at this stage to incorporate the mechanism of bandgap absorption and ensure uniformity in the n-p-n regions to prevent interference from other fields.

The structures described above exhibit differences from typical n-p structures. For certain wavelengths, the following observations are made:

(a) The absorption depths of waves at 900 and 830 nm are 32.68 and 15.46 μm, respectively, which exceed the base width of 5.8 μm. Consequently, variations in x_m lead to insignificant changes in the number of quanta (resulting in a small slope of the absorption curve), and the response of the photocurrent change to wavelength may be erroneous.

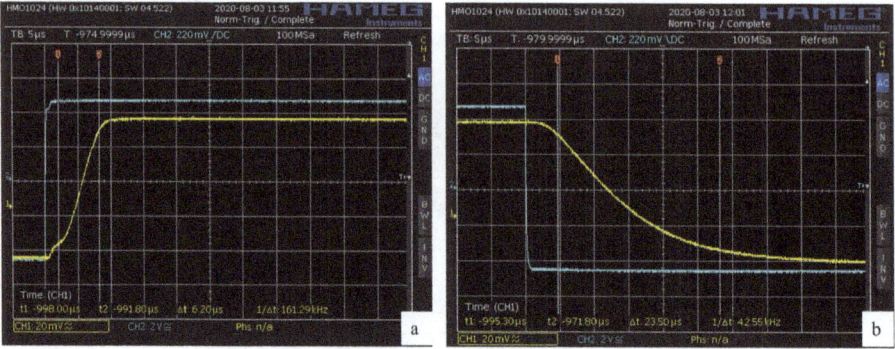

Figure 8.11: (a) The rise and **(b)** drop of the photosignal pulse for the Si2-n-p-n samples.

(b) For wavelengths of 700 and 600 nm, the absorption depths are 5.26 and 2.42 μm, respectively, which are comparable to the base width. Consequently, with variations in x_m, the intensity modulation efficiency improves significantly, resulting in a more accurate response to wavelength (as depicted in Figure 8.12a and b). At 600 nm, spectral lines of 617 and 600 nm are obtained with a radiation power of 0.34 μW, while at 700 nm, the spectral line is 681 nm with a radiation power of 7.5 μW. Hence, to achieve precise results, the absorption depth of the wave must be comparable to or less than the base depth.

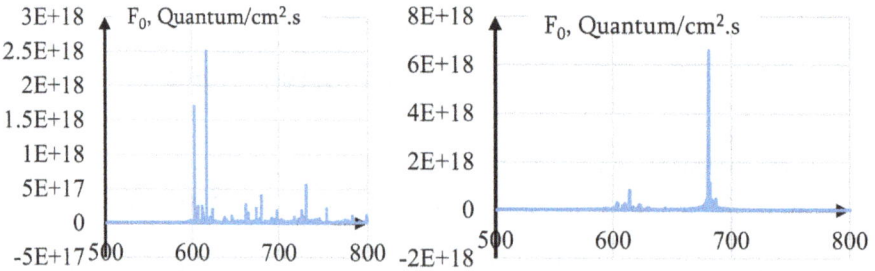

Figure 8.12: The spectral distribution of the intensity of the wave with the length of **(a)** 600 nm and **(b)** 700 nm.

The spectral distribution of radiation intensity is derived from experimental data and the power of the incident radiation on samples Si2-n-p-n. It is evident that starting from a wavelength of 460 nm (Figure 8.13, curve 1), the photocurrent and the corresponding intensity surpass those obtained from radiation absorption when the external quantum efficiency is 1 (Figure 8.13, curve 2). This discrepancy can only be explained by the amplification of the photocurrent. The spectral distributions of current photosensitivity remain identical for both signs of the bias voltage applied to the photodetector (Figure 8.14, curves 1–4 and 6–8). Such behavior can be elucidated as follows: The applied positive voltage primarily affects the reverse-biased near-surface n-p junction, and the public current depends on the reverse current of that junction. When the samples are irradiated, radiation is absorbed in both junctions.

At this juncture, the radiation that reaches the rear junction is absorbed, leading to a decrease in the barrier height of the forward-biased rear junction due to photo-generated carriers compensating for volume charges. Consequently, when irradiated, the rear barrier opens at lower voltages, allowing electrons to pass through and be injected into the base. These electrons then migrate toward the positive electrode in the n^+ region, thereby increasing the photocurrent of the near-surface barrier. The photocurrent increases with higher blocking voltage and greater absorption of quanta in the region of the rear junction.

This phenomenon occurs as the wavelength extends from 350 to 590 nm, corresponding to an increase in absorption depth or the number of quanta reaching the rear junction (Figure 8.14, curves 6–8). The peak of the spectral current photosensitivity lies in the wavelength range of 590 nm. As the wavelength further extends, the absorption depth increases while the slope of the absorption curve decreases. This results in a reduction in the number of absorbed quanta in the region of the rear junction, leading to decreased photosensitivity. In the bandgap absorption region, injection is low, and the current photosensitivity measures 9 A/W, slightly surpassing the sensitivity of the best photodiodes on the market at 0.7 A/W.

A similar trend is observed with the reverse-biased rear p-n junction, with higher current photosensitivity (Figure 8.14, curves 1–4) compared to the reversed polarity (Figure 8.14, curves 6–8). Considering that the near-surface n^+ layer has more free electrons ($n^+ \sim 10^{18}$ cm^{-3}) than the rear n^+ layer ($n^+ \sim 5 \times 10^{17}$ cm^{-3}), and the number of absorbed quanta in the near-surface junction is noticeably higher than that in the rear junction, the decrease in the height of the forward-biased near-surface barrier exceeds that of the forward-biased rear barrier.

This implies that the injection of photoelectrons from the surface to the rear n^+ layer is greater, along with the amplification of the photocurrent injection (Figure 8.14, curves 1–4). For both polarities, the spectral maximum of the current spectral photosensitivity is in the wavelength region of 590 nm. At this wavelength, the absorption depth most effectively combines with the structural parameters of the investigated samples. In the absence of bias voltage (Figure 8.14, curve 5), injection is also absent. The short-circuit photocurrent or open-circuit photovoltage have small values, approximately 2 μA and 0.022 V, respectively, corresponding to the difference in the heights of the oppositely directed potential barriers.

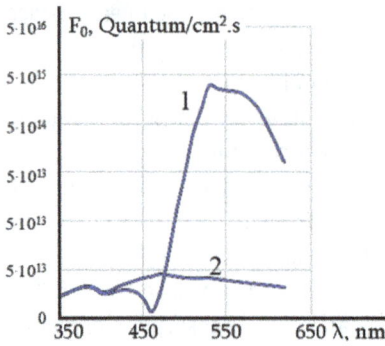

Figure 8.13: The spectral distribution of the radiation intensity for the Si2-n-p-n samples.

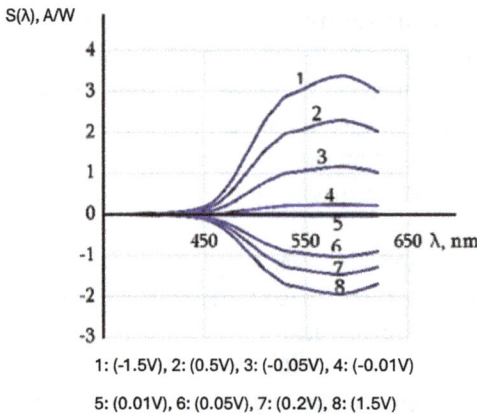

1: (-1.5V), 2: (0.5V), 3: (-0.05V), 4: (-0.01V)

5: (0.01V), 6: (0.05V), 7: (0.2V), 8: (1.5V)

Figure 8.14: The spectral distribution of the current photosensitivity for the Si2-n-p-n samples.

8.6.4 Mechanism of the selective spectral sensitivity

To achieve selective sensitivity, it is crucial to establish the real dependence of the change in x_m on the bias voltage, irrespective of its polarity. This entails a uniform change in $x_m - x_0$ and $d - x_m$ at the expense of each other. This can be validated through the distribution of the number of absorbed quanta of a given wave between the depleted regions of the two barriers and the corresponding relation of the opposite photocurrents. In such a scenario, depending on the depth of x_m or the bias voltage, the photocurrent sign may change. The wider the voltage range at which the sign change occurs, the larger the change range of x_m and the spectral range of the photocurrent sign change. This behavior is observed in all samples but to varying extents.

In samples exhibiting injection amplification of the photocurrent, such as Si-n-*p*-*n* and Si2-n-p-n, the voltage change range is insignificantly small. This suggests that the barrier widths do not change at the expense of each other. The injection of the photocarriers through the forward-biased barrier probably reduces it even at low voltages. Consequently, the current is primarily determined by the reverse-biased barrier and does not change its sign.

In silicide-n-p, a different scenario unfolds. Here, injection amplification of the photocurrent is absent, and a sign change is observed over a relatively broad range of wavelengths (630–790 nm). The dependence of the inversion point on the bias voltage follows a linear pattern (Figure 8.15). When the voltage is positive, the surface barrier is reverse-biased, and with an increase in voltage, the inversion point shifts toward longer wavelengths. Conversely, when the voltage polarity changes, the inversion point moves toward shorter wavelengths (Figures 8.16 and 8.17). In this case, the long-wave photocurrent is predominantly created by the rear barrier, forming the spectral maximum in the region of the bandgap absorption of Si. On the other hand, the short-wave photocurrent, created by the surface barrier, forms the maximum at shorter wavelengths. The spectral minimum arises between these maxima due to the compensation of the oppositely directed photocurrents. The compensation factor determines the position of the shortwave spectral maximum (Figure 8.16).

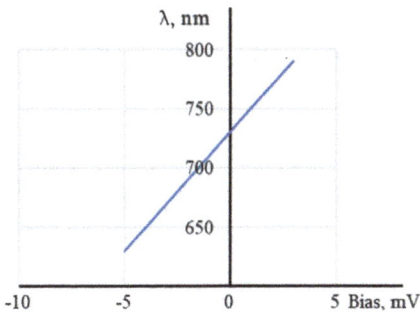

Figure 8.15: The dependence of the inversion points of the shortwave photocurrent for the silicide-n-p samples.

According to the experimental data, the shortwave change in sign does not conform to the linear relationship with the bias voltage. At high voltages on the photodetector, there is an absence of sign change in the spectral photocurrent (Figure 8.18). At a fixed voltage, the effective absorption is low due to the presence of surface recombination centers. As the wavelength increases, so does the absorption depth. The spectral photocurrent passing through the maximum decreases when influenced by the compensating rear opposite photocurrent (Figure 8.18). Thus, the linear change in the inversion point of the spectral photocurrent occurs at bias voltages of –5 to 3 mV (Figures 8.15 and 8.16).

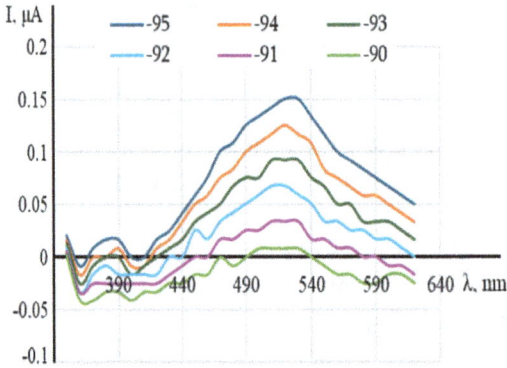

Figure 8.16: The dependence of the spectral distribution on the bias voltage for the silicide-n-p samples.

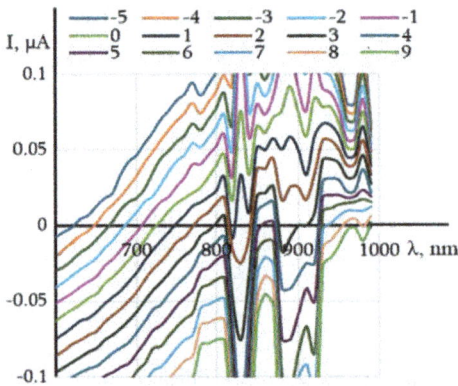

Figure 8.17: The dependence of the spectral distribution of the longwave photocurrent for the silicide-n-p samples.

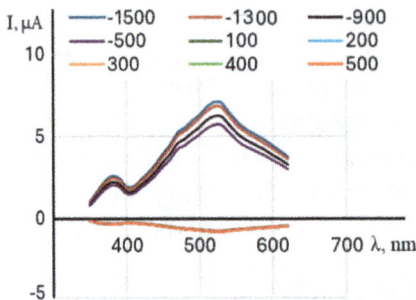

Figure 8.18: The dependence of the spectral distribution of the shortwave photocurrent for the silicide-n-p samples.

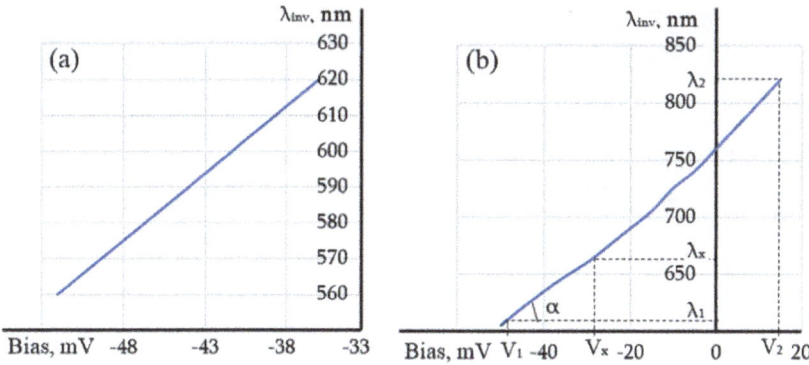

Figure 8.19: The dependence of the inversion points of the spectral photocurrent on the bias voltage for the Si1-n-p-n samples for **(a)** the shortwave and **(b)** the longwave regions.

8.6.5 Injection amplification of the photocurrent

In samples Si1-n-p-n, injection amplification is absent. However, two sufficiently deep opposite n-p and p-n junctions (as depicted in Figure 8.2) lead to a linear dependence of the inversion point of the spectral photocurrent. This occurs in the wavelength range of 560–620 nm (with a voltage change from −33 to −55 mV) and 600–830 nm (with a voltage change from 15 to −50 mV) (Figures 8.19a and 8.20, as well as Figures 8.19b and 8.21). Consequently, in the absence of injection of photocarriers from the out-of-the-base low-resistance regions, changes in voltage lead to alterations in the width of the depleted regions, and consequently, the inversion of the spectral photocurrent sign.

Obviously, the unknown wavelength of absorbed radiation can be determined based on the value tg α in the linear region of the dependence of the inversion point on the bias voltage (Figure 8.19b) using the following formula:

$$\lambda_x = \frac{V_x(\lambda_2 - \lambda_1) + \lambda_1 V_2 - \lambda_2 V_1}{V_2 - V_1} \tag{8.16}$$

Figure 8.22 illustrates the relationship between the position of x_m and the bias voltage, along with the range within which the inversion point changes. In eq. (8.6), x_m is determined by equating the current to zero and utilizing the values of L_p, d, and α. It is essential to incorporate data on the absorption coefficient corresponding to the wavelengths at the junction point.

As depicted in Figure 8.22, the constructed curve may be regarded as linear within the bounds of measurement error, affirming the linear dependency of $x_m(V)$ as described in eq. (8.3). Notably, it is the voltage range of −0.05 to + 0.015 V that actively modulates the position of x_m in the Si1-n-p-n samples. When formulating the corresponding algorithm, the section of the current-voltage characteristic within the

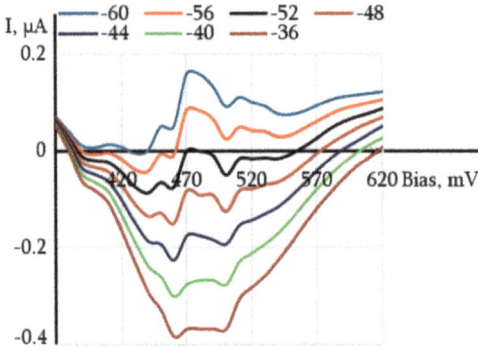

Figure 8.20: The dependence of the spectral distribution of the shortwave photocurrent for the Si1-n-p-n samples.

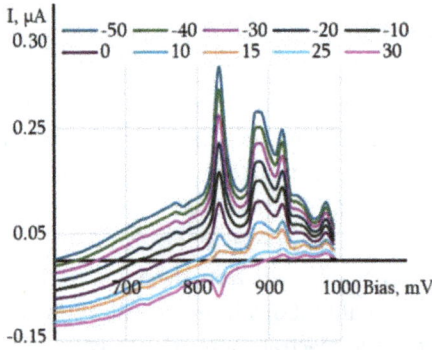

Figure 8.21: The dependence of the spectral distribution of the longwave photocurrent for the Si1-n-p-n samples.

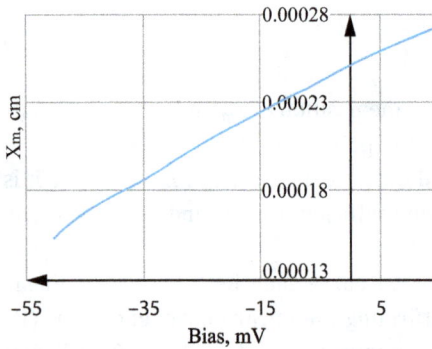

Figure 8.22: The dependence of the position of x_m on the bias voltage and the range within which the inversion point changes.

voltage range wherein the change in spectral photocurrent sign occurs can be leveraged to ascertain the radiation wavelength.

The investigation outlined above offers a valuable approach to identifying the source of any unknown wavelength in a spectrum, particularly those originating from substances in a transparent aqueous medium. Given the increasing relevance of emergent contaminants, there is a growing need for characterization tools that can effectively address unknowns rather than just known substances. This method provides a versatile and practical means of determining the origin of unidentified wavelengths, thereby enhancing our ability to detect and analyze emergent contaminants and other novel substances in various environmental and industrial contexts.

8.7 Summary

Overall, the importance of addressing emergent contaminants in transparent media cannot be overstated, as their presence poses multifaceted risks to both human and environmental health. Efforts to mitigate these risks require collaboration among scientists, policymakers, industries, and communities to develop effective strategies for monitoring, managing, and reducing the impact of these pollutants. The investigation above highlights that sufficiently long and intense waves reaching the rear junction generate photocarriers, leading to a reduction in the height of the rear potential barrier. This process facilitates electron injection through the barrier into the base, resulting in the amplification of the spectral photocurrent. Additionally, the maximum of spectral current photosensitivity coincides with the wavelength (Figure 8.6a, 590 nm), where the highest absorption of quanta in the region of the rear junction occurs, alongside a decrease in the potential barrier. The compensating junctions in Si structures cause a shift of the main maximum of spectral photosensitivity from the bandgap (~850 nm) to the shortwave region (~590 nm and 530 nm, respectively, for Si2-n-p-n and Si1-n-p-n samples). By narrowing the base (below 5.8 μm) and adjusting the impurity concentration in the base to couple the barriers, it becomes feasible to create shorter-wave Si photodetectors (with a maximum of spectral sensitivity <530 nm) possessing high sensitivity for detecting weak optical signals. Moreover, in samples lacking injection amplification of the photocurrent, an inversion of spectral photocurrent sign occurs. This inversion point linearly depends on the bias voltage, offering a method to identify unknown wavelengths. Thus, experimental samples have been examined, demonstrating the feasibility of achieving spectral-selective sensitivity by altering potential barriers in the base through external voltage manipulation.

Chapter 9
Spectral sensitivity

9.1 Introduction

Spectral sensitivity in photodetectors refers to how responsive the photodetector is to different wavelengths of light. Different types of photodetectors have varying spectral sensitivities, meaning they are more sensitive to certain wavelengths of light than others. Various devices, such as photodiodes, photomultiplier tubes, avalanche photodiodes, charge-coupled devices, CMOS sensors, and InGaAs photodiodes, display unique spectral sensitivities as per design considerations. For the devices under consideration, characteristics of an oppositely placed double potential barrier (OPDPB) photodetector structure are considered here under longitudinal illumination. The voltage range at which the inversion of the spectral photocurrent sign occurs is determined when the depleted regions join. It is shown that this range is commensurate with the difference in the heights of the potential barriers. The near-surface barrier has the advantage of quanta absorption and creates a shortwave spectral maximum of the photocurrent when counteracted by the rear barrier in the shortwave part of the spectrum. In the longwave part, the advantage of the absorption goes over to the rear barrier, and a longwave spectral maximum of the photocurrent is created. A spectral minimum appears between the maxima, which corresponds to the maximum compensation of the photocurrents of the oppositely directed barriers. The analyzed structure facilitates the selective detection of individual waves and their respective intensities. These findings herald a promising spectrophotometric future for double-barrier structures.

9.2 Selective spectral sensitivity of oppositely placed double-barrier structures

The operation of optical monitoring systems is based on the quantitative and qualitative assessment of the composition of optically transparent materials and the output optical response (Thomas & Burgess, 2017; Worsfold & Zagatto, 2017). The optical systems are intended to be applied in the sphere of agriculture, the mining industry, polymer production, veterinary medicine, food and drink industry, environmental protection against pollutants and radiations, clinical diagnostics, drug monitoring systems, fermentation systems, wastewater management systems and so on. Thus, further improvements of the systems aimed at making them smaller, cheaper, more accurate, and more sensitive, especially for the real-time remote monitoring of harmful substances, are crucial (Albert et al., 2012; Bui & Hauser, 2015; Technavio, 2020).

Modern monitoring systems, while effective, suffer from drawbacks such as high costs and lack of portability due to their bulky dimensions. These systems typically

https://doi.org/10.1515/9783111428024-010

comprise optical diffraction grids, light filters, prisms, and precise mechanical rotation nodes (Thomas & Burgess, 2017; Worsfold & Zagatto, 2017). Consequently, their utility in nonlaboratory settings, especially field applications, is limited. To address these challenges, employing spectrum-sensitive and miniature semiconductor photodetectors in monitoring systems instead of intricate and costly mechanical-optical components proves to be the most viable solution.

Currently, the international scientific community is actively engaged in advancing semiconductor photodetectors for monitoring systems, encompassing traditional materials like Si, Ge, A_3B_5 compounds, as well as thin-film and 2D materials such as graphene, perovskite, and Si carbide (Fang et al., 2019; Huo & Konstantatos, 2018; D. Liu et al., 2016; Z. Sun & Chang, 2014; Wu et al., 2018; Xu & Lin, 2020; T. Zhang et al., 2019). Specialized schematic and design-technological solutions are employed to ensure the spectral selectivity of these photodetectors (Chen et al., 2021; Hu et al., 2018; Seymour et al., 2014b; Z. Wang et al., 2019; Yin et al., 2018; Z. Zhang et al., 2011). These solutions often entail multilayer structures or several active cascade layers with varying substrate thicknesses, resulting in different degrees of photoconductivity based on the depth of wave penetration. The spectral intensity distribution is then derived through mathematical processing of measurement outcomes. However, achieving high-precision registration with these structures necessitates identical absorption conditions and the fabrication of nano-precise multilayer structures. The complex manufacturing process and the inability to control spectral sensitivity via external voltage pose challenges to their fabrication and utilization.

This study explores a novel approach and, based on it, proposes a photodetector structure utilizing an OPDPB. This structure enables the recording of the spectral composition of integral radiation emitted from the optically transparent medium under examination within the wavelength range of 300–1,000 nm. Leveraging the developed algorithm, this structure allows for superior spatial separation of information embedded in waves compared to previous works under conditions of longitudinal absorption of electromagnetic radiation. Furthermore, the algorithm facilitates the extraction of the spectral distribution of the intensity of these waves, enabling selective recording of the spectral composition of the photosignal originating from the test substance. This chapter presents the comprehensive findings of our investigation into the characteristics of OPDPB photodetectors.

9.3 Experimental device structure

Figure 9.1 illustrates a cross section of the Si OPDPB photodetector structure. The structure features a high-resistivity base region with p-type conductivity positioned between the low-resistivity n^+ regions, boasting a thickness of 5.5 μm. Space charge regions envelop the base on both sides, resembling thin n-p junctions due to their significantly narrower width compared to the electron and hole diffusion lengths. Consequently, charge carriers traverse the space charge layer without recombination, suggesting negligible recombination within the regions of the n-p junctions.

In the investigated samples, recombination currents exert minimal influence on the total photocurrent. This holds true for radiation absorption on impurities (all ionized at room temperature), free carriers, excitons, gratings, and tunneled carriers through barriers. These mechanisms may contribute to the longwave region of the spectrum only at low temperatures (Huo & Konstantatos, 2018).

The samples were fabricated utilizing the technological capabilities of the RD Alfa Microelectronics Laboratory (Riga, Latvia). An n-type Si substrate with (100) orientation, having a resistivity of 0.03 $\Omega \times$ cm and a thickness of 460 nm, served as the foundation. A base layer with a thickness of approximately 5.5 μm was epitaxially grown on this substrate, with a cathode depth of 0.3 μm, resulting in an effective film thickness of about 5.8 μm. The impurity concentration in the base was 2×10^{14} cm^{-3}, while in the cathode and substrate, it was 5×10^{18} cm^{-3} and 1×10^{18} cm^{-3}, respectively. Consequently, the base could be enveloped by depleted areas featuring OPDPB (Figure 9.1).

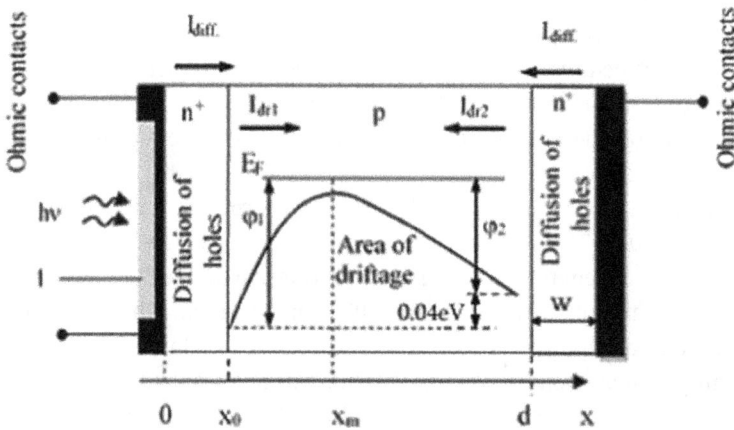

Figure 9.1: The distribution of the potential energy of holes in the valence band and the direction of the photocurrents.

The OPDPB photodetector samples underwent irradiation from various light sources featuring distinct spectral peaks. Both the total radiation and individual light sources within the wavelength range of $\lambda = 300$–1,000 nm were measured using a monochromator for this research investigation. To acquire the $I-V$ characteristics and spectral features, a Keithley 6380 Sub-Femtoamp Remote Source Meter Instrument from Cleveland, OH, USA, was employed. This instrument facilitated the application of stepwise voltage to the photodetector, with a voltage step of 1 mV. Additionally, the 6 1/2-digit Model 6430 Sub-Femtoamp Remote Source Meter Instrument was utilized to measure current with a sensitivity of 1 nA.

9.4 Results from experimental structure

The spectral dependence of the photocurrent (I_{Ph}) of OPDPB photodetector samples at various bias voltages is depicted in Figure 9.2. The spectral response exhibits the following characteristics:

(a) As the wavelength increases, the impact of the reverse photocurrent of the rear barrier gradually intensifies. Consequently, the total photocurrent passing through the shortwave maximum (in the wavelength region of 490 nm) decreases. This decrease occurs because, according to the exponential absorption law, photons penetrate deeper into the material, leading to the generated photocurrent of the rear junction counteracting the photocurrent of the near-surface barrier.

(b) At a wavelength of 600 nm, the rear photocurrent becomes comparable to the surface photocurrent. This results in maximum compensation of the opposing photocurrents, thereby determining the minimum point of the absolute photocurrent. Subsequently, within the wavelength range of 600–1,000 nm, the photocurrent predominantly governed by the rear barrier increases in absolute value. In the region of intrinsic absorption of Si, it forms a longwave maximum. At high negative voltages, this maximum is observed in the wavelength region of 830 nm. As the voltage magnitude decreases, the maximum shifts toward longer wavelengths, reaching the wavelength region of 880 nm. This shift is influenced by the sufficiently wide rear out-of-the-base region, facilitating the diffusion current of the rear junction and effective absorption of deeply penetrating longwaves.

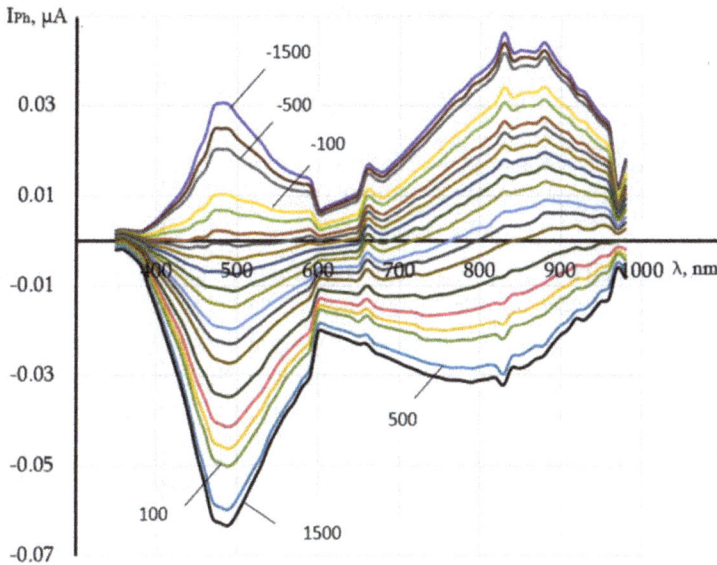

Figure 9.2: The spectral dependence of the photocurrent of OPDPB photodetector samples at various bias voltages (in mV).

(c) Under positive biases, the near-surface barrier experiences reverse biasing. As the voltage decreases, the longwave maximum (830 nm) shifts toward shorter wavelengths (730 nm). This shift occurs because the height of the near-surface barrier decreases, and its photocurrent influence remains predominant only for shorter wavelengths, where the absorption depth decreases.

Another noteworthy aspect in Figure 9.2 is the variation in the spectral photocurrent sign within the bias voltage range of −0.05 to +0.04 V. Beyond this range, the sign change does not occur due to the dominance of the photocurrent from either the reverse-biased near-surface or rear barriers. Within the specified voltage range, the shortwave photocurrent from the near-surface barrier and the longwave photocurrent from the rear barrier both exhibit prominence. For a detailed description of the change in the spectral photocurrent sign in the longwave (Figure 9.3a) and shortwave (Figure 9.3b) regions of the spectrum is shown below.

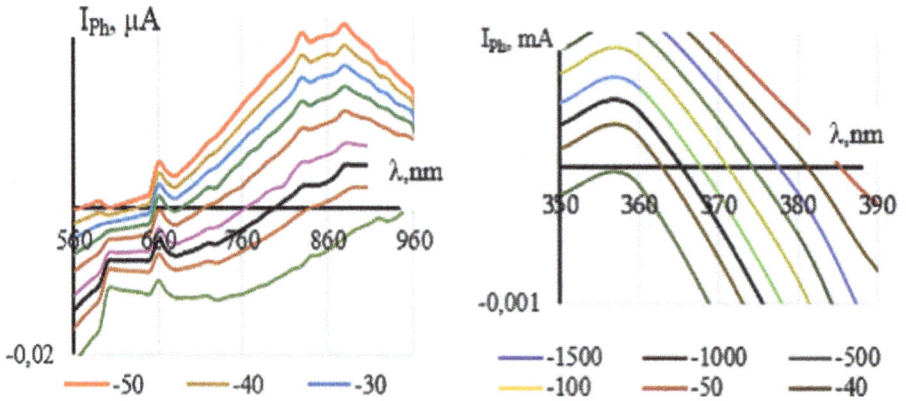

Figure 9.3: The change in the spectral photocurrent sign of OPDPB photodetector samples in the longwave (a) and shortwave (b) regions at different bias voltages (in mV).

Within the specified voltage range, as the voltage transitions from positive to negative, the height of the near-surface barrier diminishes. At −0.04 V, it becomes comparable to the height of the rear barrier. Research indicates that within the wavelength range of approximately 550–850 nm, the relationship between the inversion point of the spectral photocurrent and the bias voltage, denoted as $\lambda_{inv} = f(V)$, follows a linear trend (Figure 9.4).

The unknown wavelength is determined by the slope of that dependence by the equation:

$$\lambda_x = \frac{V_x(\lambda_2 - \lambda_1) + \lambda_1 V_2 - \lambda_2 V_1}{V_2 - V_1} \qquad (9.1)$$

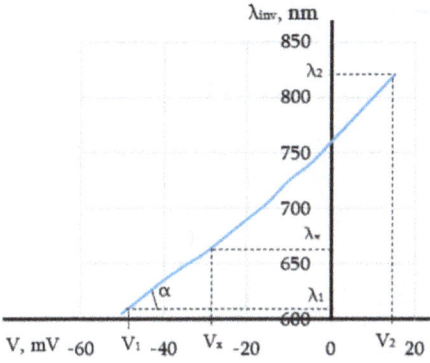

Figure 9.4: The dependence of the inversion points of the spectral photocurrent on the bias voltage.

Within the specified voltage range, the relationship between the inversion points of the spectral photocurrent and the photocurrent ($\lambda_{inv} = f(I)$) obtained by integrating the flux of radiation containing wavelengths of 550–850 nm also exhibits a linear trend (Figure 9.5). This characteristic is similarly observed in the shortwave region of the spectrum, spanning the wavelength range of 350–380 nm (Figure 9.6).

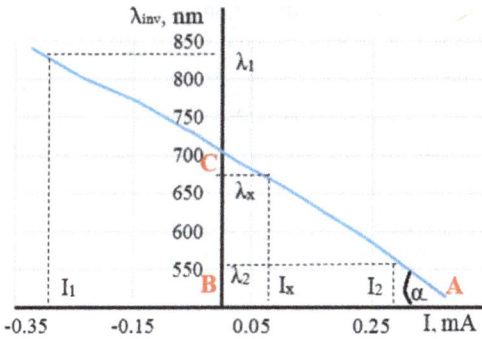

Figure 9.5: The dependence of the position of the inversion point of the spectral photocurrent on the photocurrent of the absorbed integral radiation.

It is clear that within the voltage range where the spectral photocurrent changes the sign, the unknown wavelength in the absorbed integral radiation can be determined by tg α in the linear dependence $\lambda_{inv} = f(I)$ of the inversion point of the spectral photocurrent on the total photocurrent, as depicted in triangle ABC (Figure 9.5), according to the equation:

$$\lambda_x = \frac{I_x(\lambda_2 - \lambda_1) + \lambda_1 I_2 - \lambda_2 I_1}{I_2 - I_1} \tag{9.2}$$

When there is a discrepancy in the heights of the opposing potential barriers, the photocurrent within the structure changes its direction at voltages corresponding to this differ-

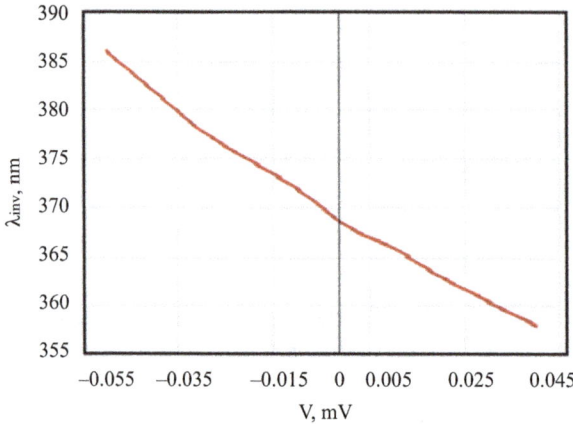

Figure 9.6: The dependence of the inversion points of the spectral photocurrent on the bias voltage in the shortwave region of the spectrum.

ence. The algorithm's logic for selective sensitivity is as follows: as the bias voltage shifts, the depleted regions of the barriers adjust, affecting the absorption fraction of penetrating quanta in both barriers. Consequently, the total photocurrent, largely influenced by the dominant intensity wave in the absorbed radiation flux at that depth, fluctuates.

In this scenario, the discrepancy in photocurrents at nearby voltage values applied to the photodetector can often be attributed to a single wave. Thus, through gradual adjustments in bias voltage using a previously established mathematical model, it becomes feasible to ascertain both the wavelength and its intensity.

The mathematical modeling of photoelectronic processes within the algorithm for spectral radiation intensity distribution incorporates an expression for determining the junction point of the depleted regions within the structure, derived from solving the Poisson equation (S. Khudaverdyan et al., 2021):

$$x_m = \frac{d - x_0}{2} - \frac{\varepsilon \varepsilon_0 (\Delta \varphi - qV)}{N_d q^2 (d - x_0)} \tag{9.3}$$

where N_d is the p-type impurity density in the base, ε is the relative permittivity of the substance, ε_0 is the permittivity of the free space, q is the electron charge, V is the bias voltage, x_0 is the thickness of the near-surface layer, d is the base width, and $\Delta \varphi = \varphi_{b1} - \varphi_{b2}$ is the difference in the heights of the potential barriers.

x_m changes linearly from x_0 up to d with the change in the bias voltage.

This suggests that the absorbed quanta are evenly redistributed among the depleted regions as the bias voltage changes, following an exponential distribution. In the context of irradiation with integral flux, the expression for photocurrent can be formulated as follows (S. Khudaverdyan et al., 2021):

$$\sum_{i,j} I_{\text{Ph}ij} = qS \sum_{i} F_{i0} \left(1 - 2e^{-a_i x_{mj}} + \frac{e^{-a_i d}}{1 + a_i w} \right) \tag{9.4}$$

where $i = 1$, 2, 3, ... changes with the change in the radiation wavelength in the integral flux and $j = 1$, 2, 3, ... changes with the change in the bias voltage, $F(\lambda_i)$ is the total flux of the incident photons with the wavelength λ_i, a_i is the absorption coefficient of the electromagnetic radiation, S is the photosensitive area, w is the width of the rear n-region, $F_0 = P_{\text{opt}}(1 - R)/Shv$ is the total flux of the incident photons per unit area, P_{opt} is the radiation power, R is the reflection coefficient, h is Planck's constant, v is the frequency of the electromagnetic radiation, and q is the electron charge.

In eq. (9.2), the width of the rear n-region w is smaller than the hole diffusion length L_p, and L_p is substituted by w (Figure 9.1). Using eq. (9.2), the photocurrent and intensity of an individual wave can be derived:

$$I = qSF_0 \left(1 - 2e^{-a_i x_{mj}} + \frac{e^{-a_i d}}{1 + a_i w} \right), \quad F_0 = I/qS \left(1 - 2e^{-a_i x_{mj}} + \frac{e^{-a_i d}}{1 + a_i w} \right) \tag{9.5}$$

When x_m reaches a sufficient value, the proportion of shortwave penetration decreases. Through incremental adjustments (in small steps) in external voltage, the absorbed quanta of each wave undergo redistribution. The predominant change in photocurrent stems from the wave exhibiting the highest intensity at depth x_m, which is then recorded. Consequently, individual waves and their intensities are registered step-by-step.

The experimental photocurrents obtained are often minute, allowing for transformation using the Maclaurin series centered around the point "x_m." This process yields an expression for the absorption coefficient

$$a = \frac{2A(x_{m2} - x_{m3}) - x_{m1} + x_{m2}}{A(x_{m2}^2 - x_{m3}^2) - x_{m1}^2 + x_{m2}^2} \tag{9.6}$$

where $A = (I_1 - I_2)/(I_2 - I_3)$ is determined from the current-voltage characteristics of the neighboring values of the photocurrents that correspond to the values of x_{m1}, x_{m2}, and x_{m3}.

Subsequently, numerical values of the absorption coefficient and the corresponding wavelengths of electromagnetic radiation in Si (Green, 2008; Smith et al., 2000) are utilized, transitioning from a to λ with the assistance of a specialized program. This facilitates the automatic acquisition of the spectral distribution of absorbed radiation intensity within the Excel environment automatically.

The powers of individual waves employed in the experiment, once intensity recalculations are performed, are depicted in Figure 9.7. The integral photocurrent resulting from the absorption of these waves is fed into the corresponding algorithm via incremental adjustments in bias voltage with a step size of 1 mV. Consequently, the experimental spectral distribution of intensity is obtained at the algorithm's output (Figure 9.7b) (S. K. Khudaverdyan et al., 2016).

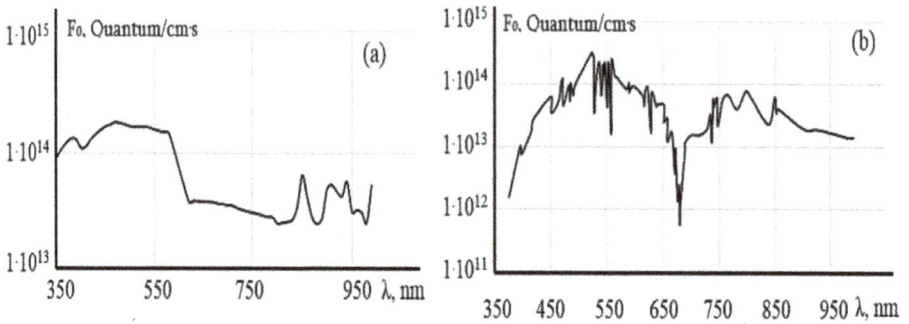

Figure 9.7: The measured (a) and the experimental (b) spectral distribution of the intensity (S. Khudaverdyan et al., 2022).

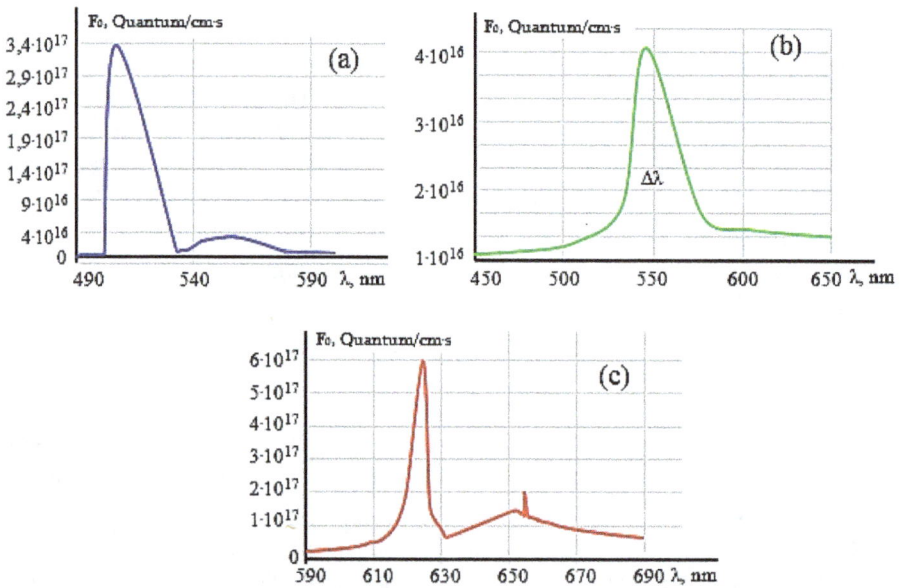

Figure 9.8: The spectral dependence of the intensity of blue (a), green (b), and red (c) LEDs (S. Khudaverdyan et al., 2022).

This algorithm was used to obtain the distribution spectra of green, blue, and red LEDs. It differs from the previously obtained spectrum of the same green radiation (Khudaverdyan S. Kh. et al., 2013) by its half-width $\Delta\lambda$, which, in this case, is close to the reference distribution of the spectral intensity of the green LED. This similarity is also observed when the blue and red radiations of the corresponding LEDs are absorbed (Figure 9.8).

The adequate similarity of these dependencies makes it possible to increase the accuracy of the determination of the spectral intensity and to realize the set target.

9.5 Summary

The spectral sensitivity of a photodetector is crucial in selecting the appropriate device for a particular application. For example, in applications where detection in the UV range is important, a photodetector with sensitivity in that range would be required. Similarly, in applications where detection in the near-infrared range is needed, a photodetector with sensitivity in that range would be chosen. In summary, the semiconductor structure with a double potential barrier and junction points of depleted regions exhibits a spectral photosensitivity distribution with maxima for both shortwave and longwave under longitudinal illumination. The discrepancy in potential barrier heights causes a change in spectral photocurrent sign, both in the absence of bias voltage and for voltages elevating the low barrier height until it matches the high barrier. Beyond this voltage range, there is no sign of change. The position where the spectral photocurrent sign changes linearly depends on the bias voltage. This results in uniform redistribution of absorbed quanta between depleted regions following an exponential law with bias voltage variation. Moreover, within the voltage range where spectral photocurrent signs invert, achieved through step-by-step voltage adjustments (1 mV increments), it becomes possible to isolate individual waves and their intensities at specific depleted region junction points. Consequently, this allows for the extraction of spectral intensity distributions.

Chapter 10
High photosensitivity in two-barrier photodetectors

10.1 Introduction

Numerous applications necessitate extracting valuable information from exceedingly faint optical signals, such as those emanating from radiation, medical imaging, industrial nondestructive testing, quantum technologies, astronomy, and various scientific measurements (Damulira et al., 2019; Fritzsche et al., 2022; Hamamatsu, 2022; Huang et al., 2020; Reimer & Cher, 2019; Sheinman et al., 2014). To discern such feeble signals, exploring photodetector designs with low-sensitivity thresholds becomes imperative, facilitating effective solutions to a myriad of fundamental scientific and technological, medical, safety, and space exploration challenges. In typical photodetectors, dark currents and noise levels constrain the detection of weak optical signals by obscuring the desired signal. Consequently, diverse technological advancements are employed in processing these photodetectors to mitigate noise levels. Fundamentally, the magnitude of dark current heavily relies on the temperature of the semiconductor structure, increasing proportionally with temperature akin to the reverse current of the p-n junction in any semiconductor device. Thus, forced cooling is sometimes utilized to diminish dark current. Moreover, dark current predominantly hinges on the bandgap width of the semiconductor, decreasing as the bandgap widens. Notably, representative dark current values for Si photodiodes at room temperature typically range in the order of nA (Huang et al., 2020; Li et al., 2016; J. Liu et al., 2021; K. Sun et al., 2018). The maximum spectral photosensitivity of Si photodiodes spans from 850 to 1,000 nm range, predominantly in the red and near-infrared regions, with a peak sensitivity of up to 0.7 A/W for Hamamatsu's long-wavelength photodiodes (Hamamatsu, 2022; Kostov et al., 2013). However, for ultraviolet radiation (~200 nm), photosensitivity diminishes by approximately a factor of 6 from its peak. Additionally, photosensitivity declines with increasing wavelength, with the optical absorption band of Si diminishing to zero beyond ~1,150 m wavelengths.

10.2 High photosensitivity in two-barrier structures

The internal amplification (Medevedev, 1970; I. M. Vikulin et al., 2008) in photodetectors exhibits a relatively slow response attributed to the final time of minority carrier escape in the base, particularly with decreasing illumination. Additionally, variations in base voltage corresponding to changes in illumination induce an additional response slowdown due to p-n junction capacitance. The operational frequency range of

https://doi.org/10.1515/9783111428024-011

phototransistors typically spans from a few megahertz to several hundred megahertz, contingent upon the characteristics of the switching circuit (S. Khudaverdyan et al., 2021; M. Wang et al., 2021).

As previously emphasized, the quest for highly sensitive photodetectors with low thresholds necessitates innovative designs and technological approaches. A study delves into the photoelectric characteristics of an a-Si structure featuring oppositely directed potential barriers under longitudinal illumination. By meticulously selecting impurity densities in the n^+-p regions, the depleted regions within the base are effectively merged, resulting in variable widths of these regions at the expense of one another. Through temperature annealing (at 950 °C for 20 min) of the guard band, the structure demonstrates exceptional spectral photosensitivity at both shortwave ($\lambda = 570$ m) and longwave ($\lambda = 850$ nm) wavelengths, with remarkable numerical values of up to 4 and 1.2 A/W, respectively.

The notable internal amplification of the photocurrent is attributed to the injection of photogenerated carriers through forward-biased potential barriers, leading to a reduction in the height of both barriers. This internal amplification, reaching up to 4 orders of magnitude, is elucidated by the ratio of the actual measured spectral photocurrent to the photocurrent calculated at a quantum yield of 1. Moreover, the samples exhibit low power levels, equivalent to noise levels lower than 10^{13} Wt × Hz$^{-1/2}$ × cm^2.

10.3 Research directions

The pursuit of highly sensitive photodetectors with minimal noise and dark currents has spurred the development of a structure featuring two oppositely directed barriers, each characterized by heights φ_1 and φ_2 in eV (Khudaverdyan et al., 2021). This configuration enables the compensation of counter photocurrents within the structure. Figure 10.1 illustrates a sectional schematic of the structure with oppositely directed potential barriers, depicting the direction of photocurrents within an n^+-p-n^+ structure.

The photodetector prototype was fabricated in the technological department of RD Alfa Microelectronics, located in Riga, Latvia. The epitaxial film (base) possessed an effective thickness of 5.8 μm. In the p-base, the impurity density was approximately 1.2×10^{14} cm^{-3}, while it was around 5×10^{18} cm^{-3} in the near-surface n^+ layer and 1×10^{18} cm^{-3} in the n^+-substrate. This configuration facilitated the coverage of the base with depleted regions from the oppositely directed potential barriers. The contact points of the depleted regions within the base were at a distance x_m. Moreover, the near-surface layer had a depth of 0.3 μm. The substrates, oriented along (100) and with a resistivity of 12.0 Ω × cm, had a thickness of 260 μm. An antireflective coating, approximately 50-nm thick of SiO_2, facilitated the transmission of UV radiation (Filipovic & Selberherr, 2022; I. Vikulin & Stafeev, 1990). The current-lead contacts were denoted by 1 and 2, as depicted in Figure 10.1. For spectral characterization, radiation was provided to the monochromator from a xenon arc lamp (in the wavelength range

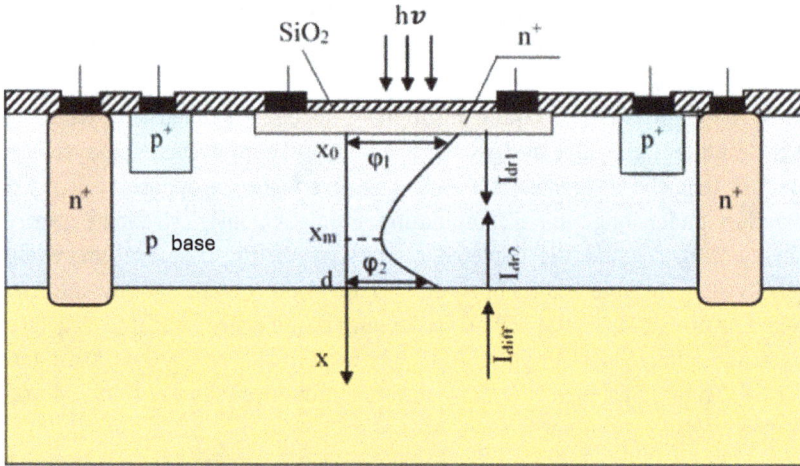

Figure 10.1: The direction of the photocurrents in the n^+-p-n^+ photodetector (reproduced with permission).

of 350–750 nm) and a halogen lamp (in the wavelength range of 700–1,000 nm), both with adjustable brightness. Current-voltage measurements were conducted using a Keithley 6340 Sub-Femto-amp Remote Source Meter from Cleveland, Ohio, USA, applying a step voltage to the photodetector with increments of 1 mV. The power of incident radiation on the photodetector was quantified utilizing Si sensor model 3A-IS, featuring a spectral range of 350–1,000 nm and a power range of 1 μW to 3 W (with a 5% error margin in wavelength).

10.4 Results of two-barrier structures

In the samples undergoing investigation, the p^+ guard ring undergoes annealing at a temperature of 950 °C for 30 min. These samples exhibit high spectral photosensitivity with two spectral peaks, as illustrated in Figure 10.2. Notably, the open-circuit voltage V_{CV1} of the sample (Figure 10.3, curve 1) measures approximately 5 mV, indicating a reduction in the calculated difference between barrier heights ($\varphi_1 - \varphi_2 = 0.4$ eV) post-annealing (Figure 10.1).

The remaining samples undergo annealing at a temperature of 1,000 °C for 30 min. Consequently, a shift in the sign of the spectral photocurrent (Filipovic & Selberherr, 2022; S. Khudaverdyan et al., 2002; S. K. Khudaverdyan, 2003; S. Kh. Khudaverdyan et al., 2014; Meliqyan et al., 2017) occurs concurrently with a change in the difference in the heights of the potential barriers, approximately 40–50 meV, comparable to the calculated value. This adjustment is evidenced by the open-circuit voltage (V_{CV}) of the sample (Figure 10.3, V_{CV2}), aligning with the calculated difference in the heights of the potential barriers.

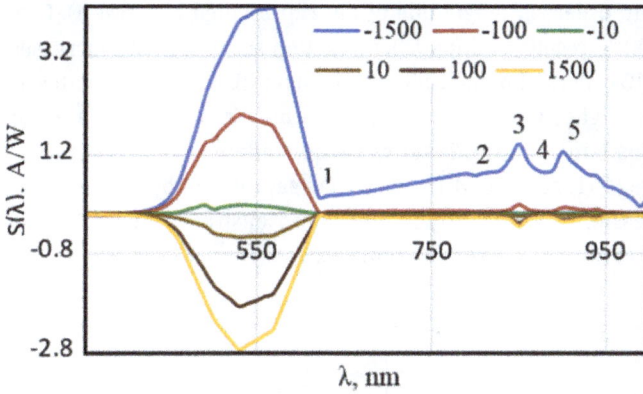

Figure 10.2: The spectral distribution of the photosensitivity of the samples upon annealing at 950 °C at different bias voltages (reproduced with permission).

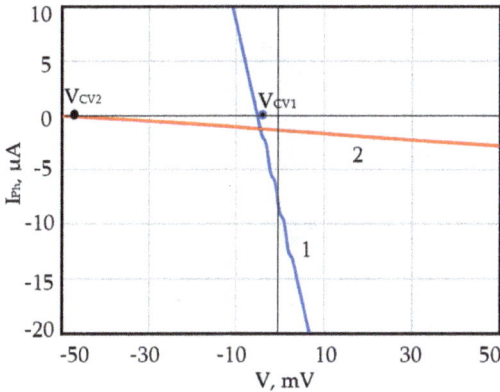

Figure 10.3: Current-voltage characteristic of the samples upon annealing at (1) 950 °C and (2) 1,000 °C (reproduced with permission).

The observed photosensitivity in both the shortwave and longwave regions is relatively low (0.03–0.04 A/W), attributed to the compensation of photocurrents from the oppositely directed barriers. However, in samples annealed at 950 °C, the current photosensitivity in the shortwave region of the spectrum (Figure 8.2, $\lambda = 570$ nm, $S_i = 4.2$ A/W) surpasses several times the spectral photosensitivity in the intrinsic region of Si (Figure 10.2, $\lambda = 850$ nm, $S_i = 1.4$ A/W). For context, high-quality Hamamatsu photodiodes typically exhibit a photosensitivity of approximately 0.7 A/W in the intrinsic absorption region of Si. Consequently, the investigated samples demonstrate anomalously high photosensitivity across both spectral regions.

The dark current-voltage characteristics (as shown in Figure 10.4) reveal that annealing at 950 °C preserves junction sharpness and maintains low levels of dark currents (up to $\pm 1{,}000\,$mV, 10 orders of magnitude of pA, represented by curve 1). Conversely, annealing at 1,000 °C results in diminished junction sharpness and increased dark currents, reaching 10 orders of magnitude of nA (curve 2). Notably, in both scenarios, dark currents are higher when the reverse-biased rear barrier has a lower height (Figure 10.4, curves 1 and 2). Additionally, the dependence of current photosensitivity on bias voltage (Figure 10.5) reveals that samples subjected to low-temperature annealing exhibit significantly higher photosensitivity (curve 1) compared to those annealed at high temperatures (curve 2).

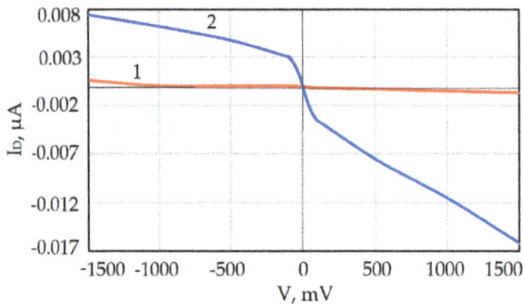

Figure 10.4: Dark current-voltage characteristics at low-temperature (950 °C, curve 1) and high-temperature (1,000 °C, curve 2) annealing (reproduced with permission).

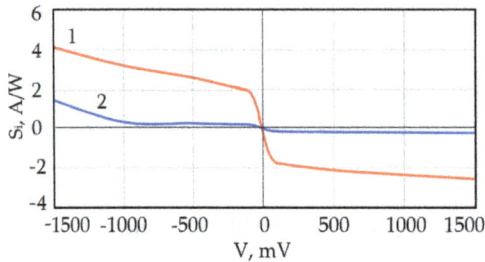

Figure 10.5: The dependence of the current photosensitivity on the bias voltage during the annealing at 950 °C (curve 1) and 1,000 °C (curve 2) (reproduced with permission).

Thus, the samples with the guard ring annealed at 950 °C had very low dark currents and high current photosensitivity.

10.5 Mechanism of high photosensitivity

It is assumed that longitudinal absorption of radiation occurs through the photosensitive surface, as depicted in Figure 10.1. Voltage changes at contacts 1 and 2 happen at the expense of each other, alongside alterations in the widths of depleted regions of the p-n junctions and the position of contact point x_m. Consequently, absorption redistributes among the depleted regions. Moreover, the n-p-n regions must maintain homogeneity to prevent noise generated by inhomogeneous fields.

The distinction between the studied structure and conventional n-p structures lies in the spectral distribution of the photocurrent. In the former, there are pronounced shortwave (560 nm) and longwave (830 nm) peaks with unusually high photosensitivity (Figure 10.2). Notably, the longwave peak falls within the intrinsic absorption region of Si.

At negative bias voltages, the surface barrier becomes forward-biased while the rear barrier is reverse-biased. The majority of the voltage is incident on the reverse-biased barrier, where waves are absorbed within the depleted layers of both potential barriers in proportion. The generated photocarriers partially diminish the heights of both barriers, creating favorable conditions for electron penetration (or injection) from the near-surface layer into the rear layer through the base. This significantly increases the concentration of minority carriers. Given that the base and near-surface layer are thinner than the electron diffusion length, their transition occurs almost without recombination.

It follows that with higher bias voltages, a greater fraction of charge carriers passes through the reverse-biased rear barrier, leading to a greater decrease in barrier height and subsequently greater photocurrent (Figure 10.2). Due to the specific ratio of impurity density in the layers, the calculated surface barrier height $\varphi 1$ is higher than that of the rear barrier $\varphi 2$ (Figure 10.1). The decrease in $\varphi 1$ is particularly noticeable with shorter waves, as most of them are absorbed in the region of the surface barrier. This accounts for an increase in spectral photocurrent in the shortwave region. As the wavelength increases, the compensating contribution of the photocurrent from the rear barrier increases, causing a decrease in spectral photocurrent passing through the structure until it reaches its minimum. Subsequently, at longer wavelengths, the rear photocurrent becomes dominant, resulting in the appearance of the longwave peak (Figure 10.2).

This observation suggests that photosensitivity is higher with the forward-biased near-surface barrier compared to the forward-biased rear barrier. This could be attributed to greater injection through the near-surface layer, where impurity density and radiation absorption are higher. Conversely, with reversed polarity of the bias voltage, a similar pattern is observed, but in reverse. The sign of the photocurrent is determined by the photocurrent of the reverse-biased barrier. Additionally, due to the exponential absorption of electromagnetic waves, a larger fraction of absorption occurs at the near-surface barrier. Consequently, photosensitivity at negative voltages (when the surface barrier is forward-biased) surpasses that at positive voltages (Figure 10.2).

With the exclusion of photocurrent stemming from background radiation and dark photocurrent resulting from the thermal generation of electron-hole pairs in the base, it can be asserted that the current traversing the structure equals the sum of dark and light currents. Under these circumstances, the threshold photosensitivity is defined by a signal-to-noise ratio of 1: $I/\sqrt{\bar{I}^2} \equiv (S_i(\lambda)P_{opt})/\sqrt{\bar{I}^2} = 1$. The radiation power is determined by the photon energy $(h\upsilon)$, the photosensitive surface (S), and the radiation intensity (F_0):

$$P_{opt} = h\upsilon \times S \times F_0 \tag{10.1}$$

Further, the RMS value of the current fluctuation is a fractional noise, which is given by

$$\bar{I}^2_{dr} = 2qI_{tot}\Delta f \tag{10.2}$$

It is the shot noise that is dominant for most semiconductor devices (Sze et al., 2021). Thus,

$$P_{threshold} = \frac{(2qI_{tot}\Delta f)^{1/2}}{S_i(\lambda)} \tag{10.3}$$

where Δf is the frequency bandpass, and $I_{tot} = I_D + I_L$ is the sum of the dark and light currents. Therefore,

$$S_i(\lambda) = I_{Ph}/P_{opt} \tag{10.4}$$

is the current photosensitivity of the structure, and P_{opt} is the power of the absorbed radiation.

With the help of eq. (10.5), we can determine the noise equivalent power (NEP) of the photodetector, with the given photosensitive surface and the unit of the bandpass, by the following expression:

$$P_{NEP} = \frac{S}{S_i(\lambda)}\sqrt{\frac{I^2}{\Delta f}} \tag{10.5}$$

Figure 10.6 displays the spectral dependence of the power acquired from experimental data, equivalent to the noise at various voltages indicated in millivolts. The photosensitive surface area of the samples is 0.021 cm^2. Particularly at a wavelength of 350 nm where photosensitivity is low, NEP exhibits a high value.

As the wavelength increases, the photosensitivity rises, leading to a decrease in NEP. Within the wavelength range of 450–570 nm, NEP remains below 2×10^{13} Wt Hz$^{-1/2}$ cm^2 at all bias voltages, and reaches its minimum at the wavelength of 550 nm. The number 2×10^{13} with units cannot be reproduced here in the text box but is shown in the actual text. This corresponds to the minimum photocurrent passing through the structure, indicative of the maximum compensation of the photocurrents from the oppositely directed potential barriers. Furthermore, with the wavelength increasing beyond 620 nm, NEP exhibits a zigzag pattern, with maxima and minima (points 1–5 in Figure 10.6) corresponding to the minima and maxima of spectral photosensitivity (points 1–5 in Figures 10.2 and 10.7) within the wavelength range of 620 –1,000 nm.

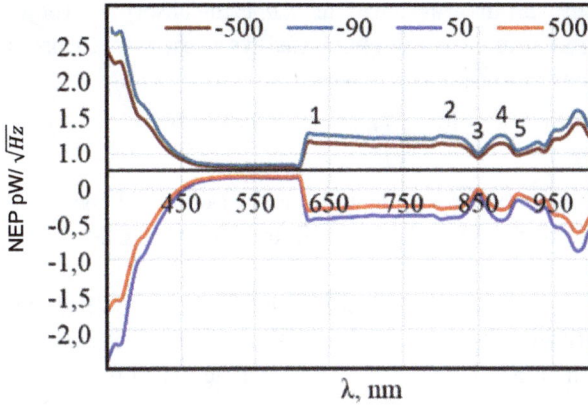

Figure 10.6: The spectral dependence of the power equivalent to the noise at different bias voltages (reproduced with permission).

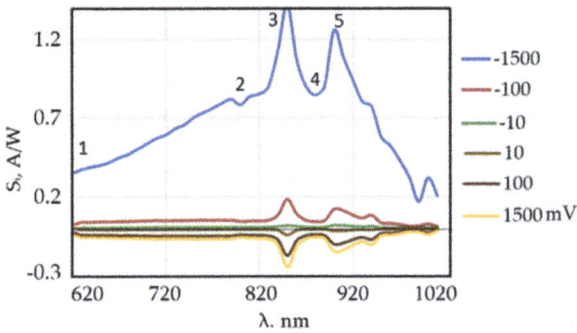

Figure 10.7: The spectral distribution of the photosensitivity of samples at different bias voltages (reproduced with permission).

With the aid of the radiation power P_{opt} utilized in the experiment and the individual monochromatic waves incident on the sample, it is feasible to compute the number of absorbed quanta with a reflection coefficient equal to unity using the expression:

$$F_0 = P_{opt}\lambda/1.24 \qquad (10.6)$$

A similar dependence is obtained with the help of the measured values of the photo-currents by the previously obtained expression (S. Khudaverdyan, et al., 2022):

$$F_0 = I/qS\left(1 - 2e^{-a_i x_{mj}} + \frac{e^{-a_i d}}{1 + a_i w}\right) \qquad (10.7)$$

where $i = 1, 2, 3, \ldots$ changes with the change in the radiation wavelength in the integral flux and $j = 1, 2, 3, \ldots$ changes with the change in the bias voltage, $F(\lambda_i)$ is the total flux of the incident photons with the wavelength of λ_i, α_i is the absorption coefficient of the electromagnetic radiation, S is the photosensitive area, w is diffusion length of minority charge carriers in the n region, where

$$F_0 = P_{opt}(1-R)/Sh\nu \tag{10.8}$$

is the total flux of the incident photons per unit area, P_{opt} is the radiation power, R is the reflection coefficient, h is Planck's constant, ν is the frequency of the electromagnetic radiation, and q is the electron charge.

Figure 10.8 illustrates the spectral ratio of the experimental (with amplification) and calculated (without amplification) photocurrents at various bias voltages, delineating the internal amplification of the photocurrent traversing the structure.

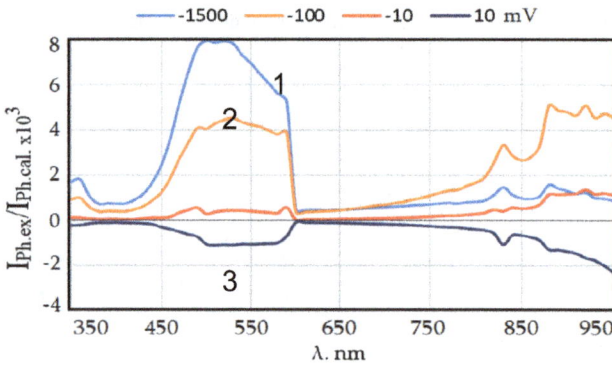

Figure 10.8: The spectral ratio of the experimental (with amplification) and calculated (without amplification) photocurrents at different bias voltages (reproduced with permission).

Figure 10.8 demonstrates that when the near-surface barrier is forward-biased, the shortwave amplification increases with higher voltage, reaching 8×10^3 times compared to the longwave amplification, which reaches 1.5×10^3 times (curve 1). As the bias voltage increases, the region of the surface barrier narrows, while the region of the rear barrier expands. This redistribution of the number of absorbed quanta results in an increase, and consequently, amplification of the photocurrent of the rear barrier and a decrease of the near-surface barrier (curve 2). Conversely, with reverse polarity, where the rear barrier is forward-biased, the opposite pattern is observed (curve 3).

10.6 Summary

The photoelectric characteristics of n^+-p-n^+ Si structures featuring oppositely directed potential barriers have been examined. The merging of depleted regions within the base is facilitated by selecting appropriate impurity densities in the n^+, p, and n^+ regions. Variations in external bias voltage applied to the sample alter the widths of the barriers, redistributing fractions of electromagnetic wave absorption between them. Their opposing photocurrents compensate each other, ensuring low dark currents (several tens of picoamperes) and photosensitivity thresholds (in the region of the shortwave maximum of spectral photosensitivity lower than 10^{13} Wt × Hz$^{-1/2}$ × cm^2).

Under longitudinal illumination, spectral distributions of photosensitivity with shortwave ($\lambda = {\sim}570$ nm) and longwave ($\lambda = {\sim}850$ nm) peaks are observed. The numerical values of photosensitivity are remarkably high, reaching up to 4 A/W. Internal amplification of the photocurrent occurs, attributed to the injection of photogenerated carriers through the forward-biased potential barrier, resulting in decreased heights of both potential barriers. This internal amplification, reaching up to 4 orders of magnitude, is evident from the ratio of actual measured spectral photocurrent to photocurrent calculated at a quantum yield equal to 1. From the device structures stated above, several conclusions can be drawn:

- Diode structures based on cadmium tellurium have been created, which show high sensitivity in the visible and X-ray spectrum.
- The implementation of the photocurrent, injection amplification, mechanism has been experimentally proven in diode structures, under the conditions of direct deflection and absorption of visible light and X-rays.
- It has been shown that the injection amplification under optical beam absorption is a consequence of modulation of the conduction of the high-resistivity underlayer and the base caused by the redistribution of the voltage applied to the diode.
- Under X-ray absorption conditions, the resistance of the base is also modulated (since the penetration depth of X-rays in cadmium telluride is 10–50 μm, at quantum energy 20 keV), and the p-n junction is located at several microns from the radiation surface.
- The photoelectric properties of n$^+$-p-n$^+$ Si structures featuring oppositely directed potential barriers have been investigated. The merging of depleted regions within the base is facilitated by selecting appropriate impurity densities in the n$^+$, p, and n$^+$ regions. Under longitudinal illumination conditions, spectral distributions of photosensitivity with shortwave ($\lambda = {\sim}570$ nm) and longwave ($\lambda = {\sim}850$ nm) peaks have been observed. Notably, the numerical values of photosensitivity are exceptionally high, reaching up to 4 A/W. Internal amplification of the photocurrent occurs within these structures. This phenomenon is attributed to the injection of photogenerated carriers through the forward-biased potential barrier, resulting in decreased heights of both potential barriers. The extent of internal amplification, reaching up to 4 orders of magnitude, is evidenced by the ratio of the actual measured spectral photocurrent to the photocurrent calculated at a quantum yield equal to 1.
- A double-barrier semiconductor photodetector p$^+$(PtSi)-n(Si)-p$^+$(Si) structure compatible with the technological cycle of IS production has been developed. Se-

lective recording of individual waves was carried out, their intensities were analyzed, and an opportunity was created to quantitatively assess the components of optically transparent objects.

– An algorithm for obtaining spectral-selective sensitivity was developed, in the process of which a transcendental equation for obtaining the exact value of the absorption coefficient was solved. Using it, the spectral distributions of blue, green, red, and white LEDs and xenon lamp rays were obtained, which repeat the standard distributions of the corresponding sources with an accuracy of 10–30 nm.

– The threshold photosensitivity and speed of the photodetector were evaluated. The corresponding 3.4×10^{-14} Wt $\times$ Hz$^{-1/2}$ and 10^{-10} s values were obtained. These are considered high indicators. Capabilities of recording very small dark currents (up to 10 pA order) and very small intensities ($\sim 10^8$ sq./cm$^2 \times$ s) and powers ($\sim 3 \times 10^{-11}$ W) were also obtained. It is extremely important in studies of weak cosmic signals.

– The resulting structural parameters in encapsulated samples provide spectral-selective sensitivity in the range of 300–1,000 nm wavelengths and open the possibility of creating new types of spectrometric devices that will exclude the optical systems in existing devices; therefore, they will be more reliable (they will not require optical correction), light, less material-intensive, cheap, fast, suitable for field conditions, and especially for monitoring large areas.

References

Adachi, S. (1985). GaAs, AlAs, and Al$_x$Ga$_{1-x}$As: Material parameters for use in research and device applications. *Journal of Applied Physics, 58*(3), R1–R29. https://doi.org/10.1063/1.336070

Albert, D. R., Todt, M. A., & Davis, H. F. (2012). A low-cost quantitative absorption spectrophotometer. *Journal of Chemical Education, 89*(11), 1432–1435. https://doi.org/10.1021/ed200829d

Aravanis, A. M., Wang, L.-P., Zhang, F., Meltzer, L. A., Mogri, M. Z., Schneider, M. B., & Deisseroth, K. (2007). An optical neural interface: *In vivo* control of rodent motor cortex with integrated fiberoptic and optogenetic technology. *Journal of Neural Engineering, 4*(3), S143–S156. https://doi.org/10.1088/1741-2560/4/3/S02

Arustamyan, V. E., Khudaverdyan, A. S., & Tsaturyan, S. K. (2010). Investigation of the selective sensitivity of two-barrier structures in the optical region of the spectrum. *Vestnik SEUA, 2*(1), 188–191.

Atvars, A., Khudaverdyan, S., Lapkis, M., & Rudenko, S. (2019). Miniature diode spectrometer design. In N. Karafolas, Z. Sodnik, & B. Cugny (Eds.), *International Conference on Space Optics – ICSO 2018* (p. 235). SPIE. https://doi.org/10.1117/12.2536155

Ball, D. W. (2006). *Field Guide to Spectroscopy*. SPIE Press.

Becker, E. M., & Farsoni, A. T. (2014). Wireless, low-cost, FPGA-based miniature gamma ray spectrometer. *Nuclear Instruments and Methods in Physics Research Section A: Accelerators, Spectrometers, Detectors and Associated Equipment, 761*, 99–104. https://doi.org/10.1016/j.nima.2014.05.096

Blank, T. V., & Gol'dberg, Yu. A. (2003). Semiconductor photoelectric converters for the ultraviolet region of the spectrum. *Semiconductors, 37*(9), 999–1030. https://doi.org/10.1134/1.1610111

Bosio, A. (2023). CdTe-based photodetectors and solar cells. In G. Korotcenkov (Ed.), *Handbook of II-VI Semiconductor-Based Sensors and Radiation Detectors* (pp. 205–230). Springer International Publishing. https://doi.org/10.1007/978-3-031-20510-1_9

Brčeski, I., & Vaseashta, A. (2021). Environmental forensic tools for water resources. In A. Vaseashta & C. Maftei (Eds.), *Water Safety, Security and Sustainability* (pp. 333–370). Springer International Publishing. https://doi.org/10.1007/978-3-030-76008-3_15

Bui, D. A., & Hauser, P. C. (2015). Analytical devices based on light-emitting diodes – A review of the state-of-the-art. *Analytica Chimica Acta, 853*, 46–58. https://doi.org/10.1016/j.aca.2014.09.044

Campbell, J. C., Li, T., Wang, S., Beck, A. L., Collins, C. J., Yang, B., Lambert, D. J. H., Dupuis, R. D., Carrano, J. C., Schurman, M. J., & Ferguson, I. T. (2000). *AlGaN/GaN Ultraviolet Photodetectors* (E. W. Taylor Ed.; p. 124). https://doi.org/10.1117/12.405335

Chen, B., Chen, Y., & Deng, Z. (2021). Recent advances in high speed photodetectors for eSWIR/MWIR/LWIR applications. *Photonics, 8*(1), 14. https://doi.org/10.3390/photonics8010014

Chernyaev, V. N. (1977). *Technology of Production of Integrated Circuits and Microprocessors* (Radio i Svyaz, Moscow, 1977) [in Russian].

Cola, A., Farella, I., Anni, M., & Martucci, M. C. (2012). Charge Transients by Variable Wavelength Optical Pulses in CdTe Nuclear Detectors. *IEEE Transactions on Nuclear Science, 59*(4), 1569–1574. doi: 10.1109/TNS.2012.2194509.

Damulira, E., Yusoff, M. N. S., Omar, A. F., & Mohd Taib, N. H. (2019). A review: photonic devices used for dosimetry in medical radiation. *Sensors, 19*(10), 2226. https://doi.org/10.3390/s19102226

Egmond, H. P. van, & Jonker, M. A., Food and Agriculture Organization of the United Nations. (Eds.). (2004). *Worldwide Regulations for Mycotoxins in Food and Feed in 2003*. Food and Agriculture Organization of the United Nations.

Fang, Y., Armin, A., Meredith, P., & Huang, J. (2019). Accurate characterization of next-generation thin-film photodetectors. *Nature Photonics, 13*(1), 1–4. https://doi.org/10.1038/s41566-018-0288-z

Filipovic, L., & Selberherr, S. (2022). Application of two-dimensional materials towards CMOS-integrated gas sensors. *Nanomaterials, 12*(20), 3651. https://doi.org/10.3390/nano12203651

https://doi.org/10.1515/9783111428024-012

Fritzsche, S., Jaenisch, G.-R., Pavasarytė, L., & Funk, A. (2022). XCT and DLW: Synergies of two techniques at sub-micrometer resolution. *Applied Sciences, 12*(20), 10488. https://doi.org/10.3390/app122010488

G.E. (2013). *G.E Healthcare Life Sciences: Spectrophotometry Handbook* (pp. 5–11). GE.

Gergel, V. A., Lependin, A. V., Tishin, Yu. I., Vanyushin, I. V., & Zimoglyad, V. A. (2006). *Boron Distribution Profiling in Asymmetrical N$^+$-p Silicon Photodiodes and New Creation Concept of Selectively Sensitive Photoelements for Megapixel Color Photoreceivers* (K. A. Valiev & A. A. Orlikovsky, Eds.; pp. 62600C-62600C – 8). https://doi.org/10.1117/12.677027

Green, M. A., & Keevers, M. J. (1995). Optical properties of intrinsic silicon at 300 K. *Progress in Photovoltaics: Research and Applications, 3*(3), 189–192. https://doi.org/10.1002/pip.4670030303

Green, M. A. (2008). Self-consistent optical parameters of intrinsic silicon at 300K including temperature coefficients. *Solar Energy Materials and Solar Cells, 92*(11), 1305–1310. https://doi.org/10.1016/j.solmat.2008.06.009

Grigoriev, I. S., & Meĭlikhov, E. Z. (Eds.). (1997). *Handbook of Physical Quantities*. CRC Press.

Grigoryan, G. E., Grigoryan, V. V., & Khudaverdyan, S. Kh. (1997). M-P/P-M structures with photocurrent sign inversion. *Proceedings of NAS and SEUA RA, 1*(2), 143–147.

Grigoryan, K., Badalyan, Sargsyan, G.M. (2008). *Antibacterial, antifungal activity of Propolis of Apis melifera from Sisian region, Armenia*, Honeydew Honey Symposium of the Apimondia International Honey Commission on the Black Sea Cost in Bulgaria, Tsarevo, Aug. 1–3, pp. 50–54.

Hamamatsu. (2022). *Compact spectrometers with built-in Hamamatsu image sensor, optical element, etc. Mini-spectrometer* (KACC0002E28 Dec. 2023 DN) [dataset]. https://www.hamamatsu.com/content/dam/hamamatsu-photonics/sites/documents/99_SALES_LIBRARY/ssd/mini-spectro_kacc0002e.pdf

Hosch, W. (2023). *CCD*. https://www.britannica.com/print/article/106410

Hu, X., Liu, H., Wang, X., Zhang, X., Shan, Z., Zheng, W., Li, H., Wang, X., Zhu, X., Jiang, Y., Zhang, Q., Zhuang, X., & Pan, A. (2018). Wavelength selective photodetectors integrated on a single composition-graded semiconductor nanowire. *Advanced Optical Materials, 6*(12), 1800293. https://doi.org/10.1002/adom.201800293

Huang, S., Deng, G., Jin, X., Lu, Y., Song, G., Huang, H., Zhao, P., Zhang, C., Yao, J., Wu, Q., & Xu, J. (2020). The dark current suppression of black silicon photodetector by a lateral heterojunction. *Optical Materials, 110*, 110474. https://doi.org/10.1016/j.optmat.2020.110474

Huo, N., & Konstantatos, G. (2018). Recent progress and future prospects of 2D-based photodetectors. *Advanced Materials, 30*(51), 1801164. https://doi.org/10.1002/adma.201801164

Ivanov, V. I. (1988). *Course of Dosimetry*. Energoatomizdat.

Jerman, J. H., & Clift, D. J. (1991). Miniature Fabry-Perot interferometers micromachined in silicon for use in optical fiber WDM systems. *TRANSDUCERS '91: 1991 International Conference on Solid-State Sensors and Actuators. Digest of Technical Papers*, 372–375. https://doi.org/10.1109/SENSOR.1991.148888

Jiang, P., Xia, H., He, Z., & Wang, Z. (2009). Design of a water environment monitoring system based on wireless sensor networks. *Sensors, 9*(8), 6411–6434. https://doi.org/10.3390/s90806411

Joint FAO/WHO Expert Committee on Food Additives. Meeting (68th : 2007 : Geneva, S., World Health Organization, & Food and Agriculture Organization of the United Nations. (2007). Evaluation of certain food additives and contaminants: Sixty-eighth report of the Joint FAO/WHO Expert Committee on Food Additives. *Sixty-Eighth Report of the Joint FAO/WHO Expert Committee on Food Additives*, 225.

Kalkhoran, N. M., & Namayar, F. (1997). *Multi-band Spectroscopic Photodetector Array* (USA Patent US005671914A). http://www.google.com/patents/US5671914

Khudaverdyan, S., Avetsiyan, A., Khudaverdyan, D., & Vaseashta, A. (2013). Photoelectric properties of selectively sensitive sensors for the detection of hazardous materials. In A. Vaseashta & S. Khudaverdyan (Eds.), *Advanced Sensors for Safety and Security* (pp. 183–191). Springer Netherlands. https://doi.org/10.1007/978-94-007-7003-4_15

Khudaverdyan, S., Dokholyan, J., Arustamyan, V., Khudaverdyan, A., & Clinciu, D. L. (2009). On the mechanism of spectral selective sensitivity of photonic biosensors. *Nuclear Instruments and Methods in*

Physics Research Section A: Accelerators, Spectrometers, Detectors and Associated Equipment, 610(1), 314–316. https://doi.org/10.1016/j.nima.2009.05.094

Khudaverdyan, S., Dokholyan, J. G., & Khudaverdyan, D. S. (2002). New developments in photodetection. *NIM-A 567*. Conference on "New developments in photodetection, Beaune.

Khudaverdyan, S. K. (1999). Two-barrier photodetector structures with a high-resistance layer based on n-CdTe. *Modeling, Optimization and Control, SEUA, 2*, 82–88.

Khudaverdyan, S. K. (2003). Peculiarities of the spectral distribution of the photocurrent in structures based on CdTe. *Izyvestia NAS and SEUA RA, Ser. Tech. Sci. Nauk, LVI*(1), 142–148.

Khudaverdyan, S. K., Arustamyan, V. E., Dokholyan, Z. G., & Gharibyan, K. B. (2014). On the possibility of creating injection detectors of ionizing radiation. *Information Technologies, Electronics, and Radio Engineering, SEUA, 1*(17), 48–58.

Khudaverdyan, S. K., Arustamyan, V. E., Dokholyan, Zh. D., Drobysheb, G., & Khudaverdyan, A. S. (2011). Photodetector for remote spectral analysis of informative radiation. *Bulletin of SEUA, Modeling, Optimization, Management, 1*(14), 92–98.

Khudaverdyan, S. K., Grigoyan, G. E., & Pogosyan, L. N. (1998). The creation and investigation of photoelectric features of the double-barrier structures with narrow recrystallized base. *Materials,* 242–244.

Khudaverdyan, S. K., & Khachatryan, M. K. (2019). Injection amplification of photocurrent in diode structures based on CdTe. *Bulletin of NPUA. Information Technologies, Electronics, Radio Engineering, 1,* 91–103.

Khudaverdyan, S. K., & Kocharyan, A. (2004). Photoelectrical properties of structures with a high resistance layer on the recrystallized Si base. *Journal of Applied Electromagnetism, 6*(2), 1–9.

Khudaverdyan, S. K., Stepan, T., & Vaseashta, A. (2016). Selective sensitivity sensor for explosive detection and identification. In *Meeting Security Challenges Through Data Analytics and Decision Support* (pp. 99–107). IOS. https://ebooks.iospress.nl/publication/45776

Khudaverdyan, S. Kh. (2003). Photo-detecting characteristics of double barrier structures. *Nuclear Instruments and Methods in Physics Research Section A: Accelerators, Spectrometers, Detectors and Associated Equipment, 504*(1–3), 350–353. https://doi.org/10.1016/S0168-9002(03)00768-X

Khudaverdyan, S. Kh., Dokholyan, J. G., Kocharyan, A. A., Kechiyantz, A. M., & Khudaverdyan, D. S. (2005). On functional potentiality of photodiode structures with a high-resistance layer. *Solid-State Electronics, 49*(4), 634–639. https://doi.org/10.1016/j.sse.2004.12.010

Khudaverdyan, S. Kh., Vaseashta, A., & Merhabyan, N. (2014). Double-channel optical model for spectral analysis of integral flux of radiation. *International Conference Tbilisi-Spring-2014, ARW.,* 121–122.

Khudaverdyan, S., Khachatryan, M., Khudaverdyan, D., Caturyan, S., & Vaseashta, A. (2013a). New model of spectral analysis of integral flux of radiation. In A. Vaseashta & S. Khudaverdyan (Eds.), *Advanced Sensors for Safety and Security* (pp. 261–269). Springer Netherlands. https://doi.org/10.1007/978-94-007-7003-4_23

Khudaverdyan, S., Khachatryan, M., Khudaverdyan, D., Caturyan, S., & Vaseashta, A. (2013b). New model of spectral analysis of integral flux of radiation. In A. Vaseashta & S. Khudaverdyan (Eds.), *Advanced Sensors for Safety and Security* (pp. 261–269). Springer Netherlands. https://doi.org/10.1007/978-94-007-7003-4_23

Khudaverdyan, S., Kocharyan, A., & Dokholyan, J. (2005). Photoreceiver structures with extended functional potentiality on the CdTe base. *Journal of Physics D: Applied Physics, 38*(2), 272–275. https://doi.org/10.1088/0022-3727/38/2/012

Khudaverdyan, S., Meliqyan, V., Hovhannisyan, T., Khudaverdyan, D., & Vaseashta, A. (2017). Identification and analysis of hazardous materials using optical spectroscopy. *Optics and Photonics Journal, 07*(01), 6–17. https://doi.org/10.4236/opj.2017.71002

Khudaverdyan, S., Petrosyan, O., Dokholyan, J., Khudaverdyan, D., & Tsaturyan, S. (2012). Modeling of a new type of an optoelectronic biosensor for the monitoring of the environment and the food products: Optoelectronic biosensor for monitoring environment. In A. Vaseashta, E. Braman, & P. Susmann (Eds.), *Technological Innovations in Sensing and Detection of Chemical, Biological, Radiological, Nuclear Threats and Ecological Terrorism* (pp. 179–184). Springer Netherlands. https://doi.org/10.1007/978-94-007-2488-4_17

Khudaverdyan, S., & Vaseashta, A. (Eds.). (2013). *Advanced Sensors for Safety and Security* (1st ed.). Springer Netherlands: Imprint: Springer. https://doi.org/10.1007/978-94-007-7003-4

Khudaverdyan, S., Vaseashta, A., Ayvazyan, G., Khachatryan, M., Atvars, A., Lapkis, M., & Rudenko, S. (2021). On the semiconductor spectroscopy for identification of emergent contaminants in transparent mediums. In A. Vaseashta & C. Maftei (Eds.), *Water Safety, Security and Sustainability* (pp. 663–689). Springer International Publishing. https://doi.org/10.1007/978-3-030-76008-3_29

Khudaverdyan, S., Vaseashta, A., Ayvazyan, G., Matevosyan, L., Khudaverdyan, A., Khachatryan, M., & Makaryan, E. (2022). On the selective spectral sensitivity of oppositely placed double-barrier structures. *Photonics, 9*(8), 558. https://doi.org/10.3390/photonics9080558

Kikoin, I. K. (1976). *Handbook of Table of Physical Quantities.* Atomizdat.

Kishino, K., Unlu, M. S., Chyi, J.-I., Reed, J., Arsenault, L., & Morkoc, H. (1991). Resonant cavity-enhanced (RCE) photodetectors. *IEEE Journal of Quantum Electronics, 27*(8), 2025–2034. https://doi.org/10.1109/3.83412

Koledov, L. A. (1989). *Technology and design of microcircuits, microprocessors and microassemblies.* LAN publishing (in Russian). ISBN:978-5811407668.

Komarov, F. F., Mil'chanin, O. V., Kovaleva, T. B., Solov'ev, Ya. A., Turtsevich, A. S., & Karvat, Ch. (2011). Low temperature formation of platinum silicide for Schottky diodes contact layer. *Proceedings of 9 International Conference,* 471.

Kong, S. H., Wijngaards, D. D. L., & Wolffenbuttel, R. F. (2001). Infrared micro-spectrometer based on a diffraction grating. *Sensors and Actuators A: Physical, 92*(1–3), 88–95. https://doi.org/10.1016/S0924-4247(01)00544-1

Kostov, P., Gaberl, W., & Zimmermann, H. (2013). High-speed bipolar phototransistors in a 180nm CMOS process. *Optics & Laser Technology, 46*, 6–13. https://doi.org/10.1016/j.optlastec.2012.04.011

Lewis, L., Onsongo, M., Njapau, H., Schurz-Rogers, H., Luber, G., Kieszak, S., Nyamongo, J., Backer, L., Dahiye, A. M., Misore, A., DeCock, K., & Rubin, C., The Kenya Aflatoxicosis Investigation Group. (2005). Aflatoxin contamination of commercial maize products during an outbreak of acute aflatoxicosis in Eastern and Central Kenya. *Environmental Health Perspectives, 113*(12), 1763–1767. https://doi.org/10.1289/ehp.7998

Li, C., Xue, C., Liu, Z., Cong, H., Cheng, B., Hu, Z., Guo, X., & Liu, W. (2016). High-responsivity vertical-illumination Si/Ge uni-traveling-carrier photodiodes based on silicon-on-insulator substrate. *Scientific Reports, 6*(1), 27743. https://doi.org/10.1038/srep27743

Lingg, M., Spescha, A., Haass, S. G., Carron, R., Buecheler, S., & Tiwari, A. N. (2018). Structural and electronic properties of $CdTe_{1-x}Se_x$ films and their application in solar cells. *Science and Technology of Advanced Materials, 19*(1), 683–692. https://doi.org/10.1080/14686996.2018.1497403

Liu, D., Zhou, W., & Ma, Z. (2016). Semiconductor nanomembrane-based light-emitting and photodetecting devices. *Photonics, 3*(2), 40. https://doi.org/10.3390/photonics3020040

Liu, J., Cristoloveanu, S., & Wan, J. (2021). A review on the recent progress of silicon-on-insulator-based photodetectors. *Physica Status Solidi (A), 218*(14), 2000751. https://doi.org/10.1002/pssa.202000751

Maida, J. (2016). *Technavio Releases New Report on Global Spectroscopy Market* (Technavio Research). https://www.businesswire.com/news/home/20160127005709/en/Technavio-Releases-Report-Global-Spectroscopy-Market

Medevedev, S. A. (1970). *Physics and Chemistry of the A2B6 Compound.* Mir.

Medremkomplekt Company. (2002). *Лампа дейтериевая спектральная ДДС 30 (лд2-д).* https://www.medrk.ru/shop/lampy-medicinskie/lampy-prochie/id-10060

Meliqyan, V., Khudaverdyan, D. S., Khachatryan, M. G., & Khudaverdyan, S. Kh. (2017). Selectively sensitive photodetector. *Proceedings of the Eleventh International Conference*, 48–50.

Mirkin, L. I. (2012). *Handbook of X-ray Analysis of Polycrystalline Materials*. Springer; Softcover reprint of the original 1st ed. 1964 edition (March 18, 2012).

Mogniotte, J. F., Raynaud, C., Lazar, M., Allard, B., & Planson, D. (2018). SiC lateral Schottky diode technology for integrated smart power converter. *2018 IEEE International Conference on Industrial Technology (ICIT)*, 841–846. https://doi.org/10.1109/ICIT.2018.8352287

Mohammadi, A., Baba, M., & Hirayama, H. (2009). Simulation of the carrier trapping effect in a Schottky CdTe detector. *Journal of Nuclear Science and Technology, 46*(11), 1032–1037.

Moiseev, A. A., & Ivanov, V. I. (1984). *Handbook on Dosimetry and Radiation Hygiene*. Moscow Energoatomizdat. https://www.twirpx.com/file/45203/

Monea, B., Ionete, E., Spiridon, S., Leca, A., Stanciu, A., Petre, E., & Vaseashta, A. (2017). Single wall carbon nanotubes based cryogenic temperature sensor platforms. *Sensors, 17*(9), 2071. https://doi.org/10.3390/s17092071

Nader, M. K., & Fereydoon, N. (2004). *Filterless Si-Based Ultraviolet-Selective Photodetectors* (SSC-00072; Stennis Space Center, Mississippi). Spire Corporation.

Normatov, P. I., Armstrong, R., Normatov, I. Sh., & Narzulloev, N. (2015). Monitoring extreme water factors and studying the anthropogenic load of industrial objects on water quality in the Zeravshan River basin. *Russian Meteorology and Hydrology, 40*(5), 347–354. https://doi.org/10.3103/S106837391505009X

Ocaya, R. O. (2016). Versatile CCD-based spectrometer with field programmable gate array controller core. *IET Science, Measurement & Technology, 10*(7), 719–727. https://doi.org/10.1049/iet-smt.2016.0063

Ojo, A. A., & Dharmadasa, I. M. (2019). Factors affecting electroplated semiconductor material properties: The case study of deposition temperature on cadmium telluride. *Coatings, 9*(6), 370. https://doi.org/10.3390/coatings9060370

OSRAM. (2002). *Osram 64623 HLX EVA 12V 100W M/28 GY6,35* [Specification sheet]. https://www.medrk.ru/shop/lampy-medicinskie/lampy-osram/id-23992

Owen, T. (2000). *Fundamentals of Modern UV-visible Spectroscopy: Primer*. Agilent Technologies. https://books.google.com/books?id=sz9ytAEACAAJ

Ploykrachang, K., Thong-Aram, D., Punnachaiya, S., & Baotong, S. (2011). Pocket PC-based portable gamma-ray spectrometer. *Songklanakarin Journal of Science and Technology (SJST), 33*(2), 215–219.

Popov, S. (2003). Schottky diodes for converter technology. *Electronic Components, 3*, 35–38.

Reimer, M. E., & Cher, C. (2019). The quest for a perfect single-photon source. *Nature Photonics, 13*(11), 734–736. https://doi.org/10.1038/s41566-019-0544-x

Saptari, V. (2004). *Fourier-Transform Spectroscopy Instrumentation Engineering*. SPIE Press.

Seymour, E. Ç., Freedman, D. S., Gökkavas, M., Özbay, E., Sahin, M., & Ünlü, M. S. (2014a). Improved selectivity from a wavelength addressable device for wireless stimulation of neural tissue. *Frontiers in Neuroengineering, 7*. https://doi.org/10.3389/fneng.2014.00005

Seymour, E. Ç., Freedman, D. S., Gökkavas, M., Özbay, E., Sahin, M., & Ünlü, M. S. (2014b). Improved selectivity from a wavelength addressable device for wireless stimulation of neural tissue. *Frontiers in Neuroengineering, 7*. https://doi.org/10.3389/fneng.2014.00005

Shalimova, K. V. (1985). *Physics of Semiconductors*. Energoatomizdat.

Sheinman, V., Rudnitsky, A., Toichuev, R., Eshiev, A., Abdullaeva, S., Egemkulov, T., & Zalevsky, Z. (2014). Implantable photonic devices for improved medical treatments. *Journal of Biomedical Optics, 19*(10), 108001. https://doi.org/10.1117/1.JBO.19.10.108001

Corp, S. (2009). *Molecular Spectroscopy, UV-Vis-NIR Spectroscopy* (C101-E111; UV Talk Letter). https://www.shimadzu.com/an/sites/shimadzu.com.an/files/pim/pim_document_file/journal/talk_letters/13197/jpa112002.pdf

Shimadzu Corp. (2017). *TOC Special Applications, Vol. 4* (Shimadzu Excellence in Science)).

Shklovskii, I. S. (1987). *Universe Life and Mind* (5th ed.). John Wiley & Sons Inc.

Smith, D. Y., Inokuti, M., & Karstens, W. (2000). Photoresponse of Condensed Matter Over the Entire Range of Excitation Energies: Analysis of Silicon. *Physics Essays, 13*(3), 465–472. https://doi.org/10.4006/1.3028845

Stuart, B. H. (2008). *Infrared Spectroscopy: Fundamentals and Applications*. Wiley.

Sun, K., Jung, D., Shang, C., Liu, A., Morgan, J., Zang, J., Li, Q., Klamkin, J., Bowers, J. E., & Beling, A. (2018). Low dark current III–V on silicon photodiodes by heteroepitaxy. *Optics Express, 26*(10), 13605. https://doi.org/10.1364/OE.26.013605

Sun, Z., & Chang, H. (2014). Graphene and graphene-like two-dimensional materials in photodetection: mechanisms and methodology. *ACS Nano, 8*(5), 4133–4156. https://doi.org/10.1021/nn500508c

Sze, S. M., Ng, K. K., & Li, Y. (2021). *Physics of semiconductor devices* (Fourth edition). Wiley.

Technavio. (2017). *Global Spectroscopy Market 2017–2021* (SKU: IRTNTR11573; p. 70). https://www.technavio.com/report/global-embedded-systems-global-spectroscopy-market-2017-2021

Technavio. (2020). *Global Spectrophotometer Market 2020–2024* (ID: 5157122; p. 120). https://www.researchandmarkets.com/reports/5157122/global-spectrophotometer-market-2020-2024

Thomas, O., & Burgess, C. (Eds.). (2017). *UV-visible Spectrophotometry of Water and Wastewater* (2nd ed.). Elsevier.

Trumbo, T. A., Schultz, E., Borland, M. G., & Pugh, M. E. (2013). Applied spectrophotometry: Analysis of a biochemical mixture. *Biochemistry and Molecular Biology Education, 41*(4), 242–250. https://doi.org/10.1002/bmb.20694

Vanyushin, V. I., Gergel, V. A., Zimoglyad, V. A., & Tishin, Yu. I. (2005). Adjusting the spectral response of silicon photodiodes by additional dopant implantation. *Russian Microelectronics, 34*(3), 155–159.

Vaseashta, A. (2021). Introduction to water safety, security and sustainability. In A. Vaseashta & C. Maftei (Eds.), *Water Safety, Security and Sustainability* (pp. 3–22). Springer International Publishing. https://doi.org/10.1007/978-3-030-76008-3_1

Vaseashta, A., Duca, G., & Travin, S. O. (Eds.). (2022). *Handbook of Research on Water Sciences and Society*. Engineering Science Reference.

Vaseashta, A., Gevorgyan, G., Kavaz, D., Ivanov, O., Jawaid, M., & Vasović, D. (2021). Exposome, biomonitoring, assessment and data analytics to quantify universal water quality. In A. Vaseashta & C. Maftei (Eds.), *Water Safety, Security and Sustainability* (pp. 67–114). Springer International Publishing. https://doi.org/10.1007/978-3-030-76008-3_4

Vaseashta, A., Khudaverdyan, S., & Vaseashta, A. (2013). *Advanced Sensors for Safety and Security*. Springer Netherlands Springer e-books Imprint: Springer.

Vaseashta, A., & Maftei, C. (Eds.). (2021). *Water Safety, Security and Sustainability: Threat Detection and Mitigation*. Springer International Publishing. https://doi.org/10.1007/978-3-030-76008-3

Vikulin, I. M., Kurmashev, Sh. D., & Stafeev, V. I. (2008). Injection-based photodetectors. *Semiconductors, 42*(1), 112–127. https://doi.org/10.1134/S1063782608010168

Vikulin, I., & Stafeev, V. I. (1990). The physics of semiconducting devices. *Radio and Communication4/22/2024*, 264.

Wachowiak, A., Slesazeck, S., Jordan, P., Holz, J., & Mikolajick, T. (2013). New color sensor concept based on single spectral tunable photodiode. *2013 Proceedings of the European Solid-State Device Research Conference (ESSDERC)*, 127–130. https://doi.org/10.1109/ESSDERC.2013.6818835

Wang, M., Zhang, S., Xu, Y., He, Y., Zhang, Y., Zhang, Z., & Liu, Y. (2021). Frequency response measurement of high-speed photodiodes based on a photonic sampling of an envelope-modulated microwave subcarrier. *Optics Express, 29*(7), 9836. https://doi.org/10.1364/OE.420662

Wang, Z., Yi, S., Chen, A., Zhou, M., Luk, T. S., James, A., Nogan, J., Ross, W., Joe, G., Shahsafi, A., Wang, K. X., Kats, M. A., & Yu, Z. (2019). Single-shot on-chip spectral sensors based on photonic crystal slabs. *Nature Communications, 10*(1), 1020. https://doi.org/10.1038/s41467-019-08994-5

Worsfold, P. J., & Zagatto, E. A. G. (2017). Spectrophotometry: Overview ☆. In *Reference Module in Chemistry, Molecular Sciences and Chemical Engineering* (p. B9780124095472142659). Elsevier. https://doi.org/10.1016/B978-0-12-409547-2.14265-9

Wu, Y., Li, X., Wei, Y., Gu, Y., & Zeng, H. (2018). Perovskite photodetectors with both visible-infrared dual-mode response and super-narrowband characteristics towards photo-communication encryption application. *Nanoscale, 10*(1), 359–365. https://doi.org/10.1039/C7NR06193E

Xu, Y., & Lin, Q. (2020). Photodetectors based on solution-processable semiconductors: Recent advances and perspectives. *Applied Physics Reviews, 7*(1), 011315. https://doi.org/10.1063/1.5144840

Yin, J., Tan, Z., Hong, H., Wu, J., Yuan, H., Liu, Y., Chen, C., Tan, C., Yao, F., Li, T., Chen, Y., Liu, Z., Liu, K., & Peng, H. (2018). Ultrafast and highly sensitive infrared photodetectors based on two-dimensional oxyselenide crystals. *Nature Communications, 9*(1), 3311. https://doi.org/10.1038/s41467-018-05874-2

Zhang, T., Wang, F., Zhang, P., Wang, Y., Chen, H., Li, J., Wu, J., Chen, L., Chen, Z. D., & Li, S. (2019). Low-temperature processed inorganic perovskites for flexible detectors with a broadband photoresponse. *Nanoscale, 11*(6), 2871–2877. https://doi.org/10.1039/C8NR09900F

Zhang, Z., Von Wenckstern, H., Schmidt, M., & Grundmann, M. (2011). Wavelength selective metal-semiconductor-metal photodetectors based on (Mg,Zn)O-heterostructures. *Applied Physics Letters, 99*(8), 083502. https://doi.org/10.1063/1.3628338

About the authors

Prof. Dr. Surik Khudaverdyan was born in 1952 in Javakhk, Georgia. In 1968, he graduated from Yerevan School No. 104, followed by Yerevan Polytechnic Institute in 1973, and graduate school at the Moscow Institute of Electronic Technology in 1980. He worked at IRFE NA ASSR and its Central Bureau, as a junior researcher in charge of the sector and department from 1973 to 1987. He was a candidate of physical and mathematical sciences in 1981, followed by a doctor of technical sciences in 2005. He then served as the head of the "Communication Systems" Department of the National Polytechnic Institute in 2006, and the head of the Fundamental Research Laboratory of "photoelectronic devices in optical communication systems since 2011. He is a member of the professional council for the defense of theses and the editorial board of the journal *Information Technologies, Electronics, Radio Electronics*. He managed several international and national scientific projects. According to NATO's program, as a cochair, he organized an international scientific conference within the framework of the Peace and Security Program. In 2013–2018, he was listed as the 100 effective Republic of Armenia (RA) researchers. In 2021, he was awarded with the Commemorative Medal of the Prime Minister of RA and the Gold Medal of NPUA. In 2013, he received the Laureate of the RA President's Award in the field of technical sciences and information technologies. In 2014, he received gratitude from the RA prime minister. He is the author of more than 120 scientific works and 7 inventions. He is a coeditor of the collection "Advanced Security Sensors, DOI 10.1007/978-94-007-7003-4, Springer 2013." Areas of interest include microelectronics, nanoelectronics, photonics, and optical spectral analysis.

Prof. Dr. Ashok Vaseashta is a researcher, CEO/CTO, and executive director of research at the International Clean Water Institute in Virginia, USA, and Chair Professor of Ghitu Institute of Electronic Engineering and Nanotechnologies (IEEN), Chisinau, Moldova, and he serves in several honorary positions, such as a professor at the Transylvania University of Brasov, chaired professor of nanotechnology at the Ghitu Institute of Electronic Engineering and Nanotechnologies, honorary member of the Academy of Sciences of Moldova, academician at the Euro-Mediterranean Academy of Arts and Sciences, member of CIRET in France, and senior strategic research advisor for several organizations. Inspired by nature and guided by societal necessities, he strives for technological innovations to address the global challenges of the twenty-first century. His research interests include nanotechnology – materials synthesis using novel combinatorial processes; water – contamination monitoring and remediation; critical infrastructure – safety, security, and environmental sustainability; and identifying dual-use technologies – all using the nexus of advanced technological solution platforms. He is a scholar, visionary, strategist, and dedicated futurist providing strategic leadership to promote and advance research initiatives and priorities using data-driven decisions. He received his PhD from the Virginia Polytechnic Institute and State University, Blacksburg, VA, in 1990, followed by Kobe's postdoctoral fellowship. Following his PhD, he served as a professor and researcher at Virginia Tech and Marshall University. He also served as the director of research at the Institute for Advanced Sciences Convergence and International Clean Water Institute for Norwich University Applied Research Institutes, vice-provost (rector) for research at Molecular Science Research Center in Orangeburg in South Carolina, and executive director and chair of the Institutional Review Board at a State University in New Jersey. His earlier honorary positions included a visiting professor at Riga Technical University,

https://doi.org/10.1515/9783111428024-013

Latvia; 3Nano-SAE Research Centre, University of Bucharest, Romania; and visiting scientist at the Weizmann Institute of Science, Israel. He served the U.S. Department of State in two rotations, as a strategic S&T advisor in the Bureau of International Security and Nonproliferation, Office of Weapons of Mass Destruction and Terrorism, and U.S. diplomat. His research interests span foresight, nanotechnology, environmental/ecological science, and critical infrastructure safety and security. He is the author/editor of 17 books and has published over 350 articles in scientific journals, book chapters, and conferences. He serves on the editorial boards of several prestigious journals.

Index

https://doi.org/10.1515/9783111428024-014

www.ingramcontent.com/pod-product-compliance
Lightning Source LLC
Chambersburg PA
CBHW081527220326
41598CB00036B/6360